FRACTALON 4
DOUBLE SLIT EXPERIMENT

The origin of quantum physics,
THE BIGGEST LIE IN THE HISTORY
OF MODERN PHYSICS

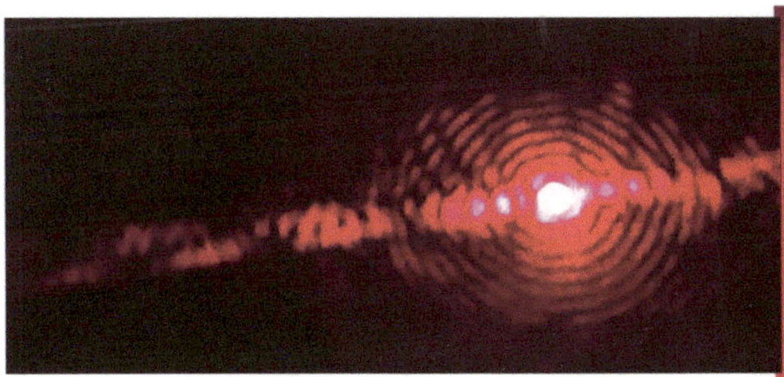

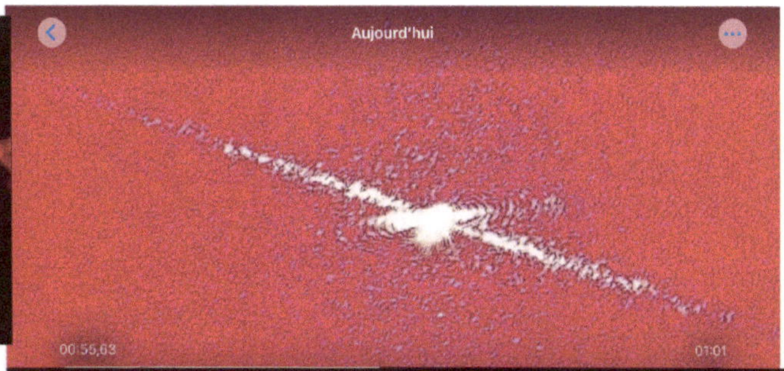

I all people wear rose-colored glasses, they will see that all the objects are pink.If all the physicists wear interferences glasses on theyr brains,they will see interferences everywhere.If you don't wear idiot glasses, you will see the truth.
TRAIAN TEREC

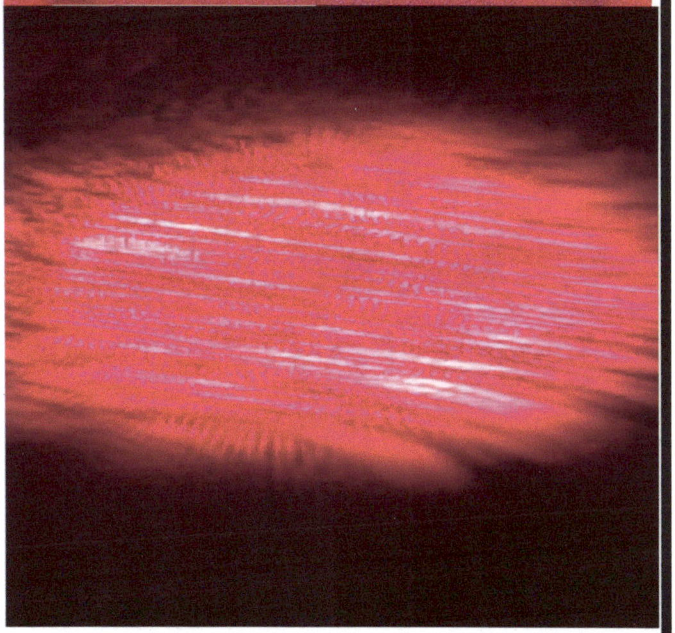

DSE=reflection of a light source like a hedgehog with spines as thick as wavelength on the surface of 3 cylindrical mirrors

What do these photos have in common?

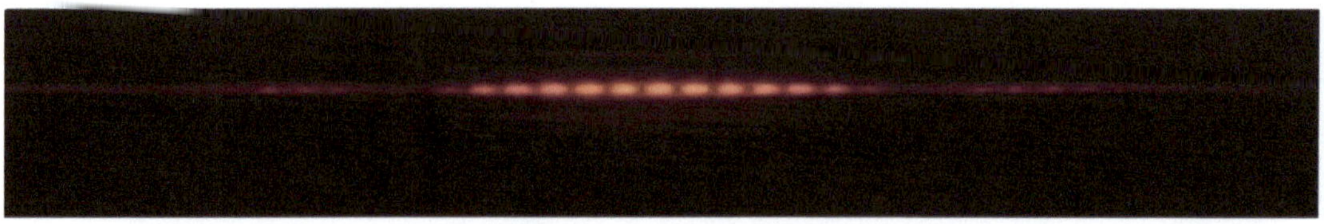

1. DSE(double slit experiment) of light

Print screen of AlphaPhoenix Youtube video about speed of electric current(+Veritassium ,ElectroBOOM)

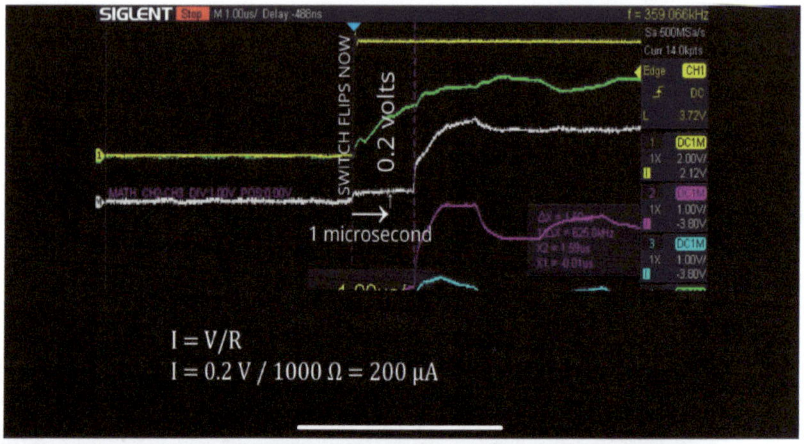

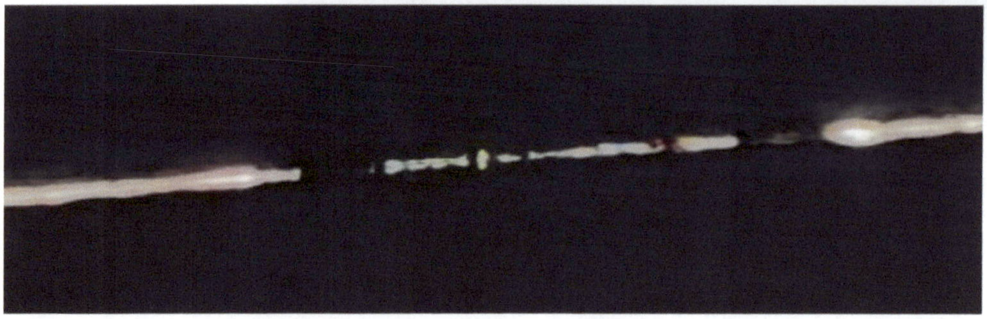

The image of a reflection on a metal miror underneath a knife

Dust filaments on a Wimshurst Machine after 48 hours of pause(Fractalon.fr video)

The reflection of white light on gas bubbles in a bottle

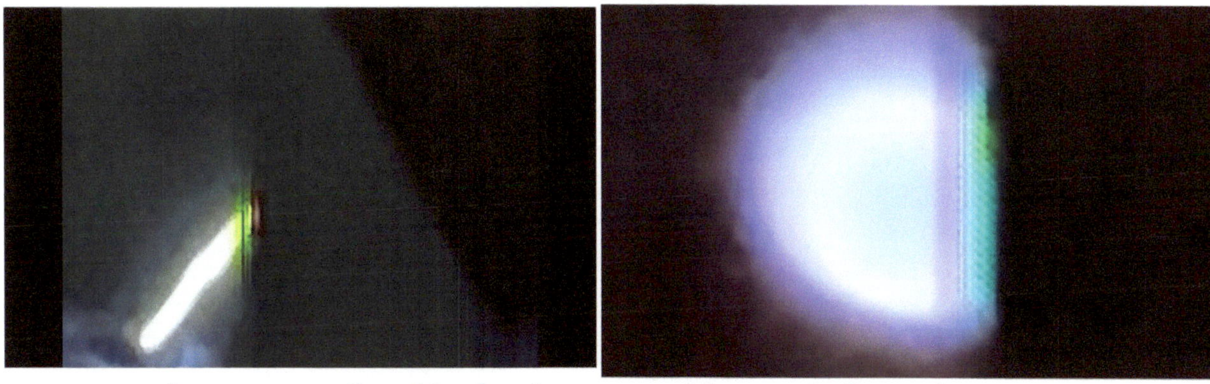

HD photos of a lighting strike with Iphone 5x camera zoom(Wimshurst machine)

A rainbow CD and the image obtained through a punched paper

To understand all these images we need to understand quantum mechanical physics, so let's start at the beggining with the double slit experiment(it has been described in detail in millions of books and videos so i won't go into details)

Material:
1.the source of light (the laser of different lenghtwave)about 1 mm width
2.the slits(2 metal wires thinner than a hair from a earphones coil- or a metal foil with two slits)
3.a Wall

A hair of mine,70000nm=0,07mm and the 2 wires (iphone 5x zoom)

The physicists consider the two wires(or slits) to be immaterial ideals, but we already know that each wavelenght is reflected at different distances from matter depending of the wavelenght

All electromagnetic waves are reflected before reaching matter at a distance that depends on their wavelenght(radio telescope)the wires of antena becomes togheter a mirror for radio lenghwave.

A knife on a METAL mirror in sun light reflection,we see all the RAINBOW depending on the width of slit, every lenghtwave need his diameter of bubble pression of fractalons to pass under the knife, so the center of the wave is at differents distances relative to the mirror(the picture has multireflected light in it like DSE but regardless of the angle chosen for the green slit we will not find the red color

In fact our light source looks like this in reality:the picture of two wires in red and green light = the cylindrical mirror represents the virtual periphery at a distance close to red wavelenght respectively green from the wire (the diameter of the mirror is the diameter of the wire + (2 x wavelenght))(image taken with iphone x5 zoom-10 cm image of a wire of 0.07mm). So is the laser image reflected in the mirror of the wires

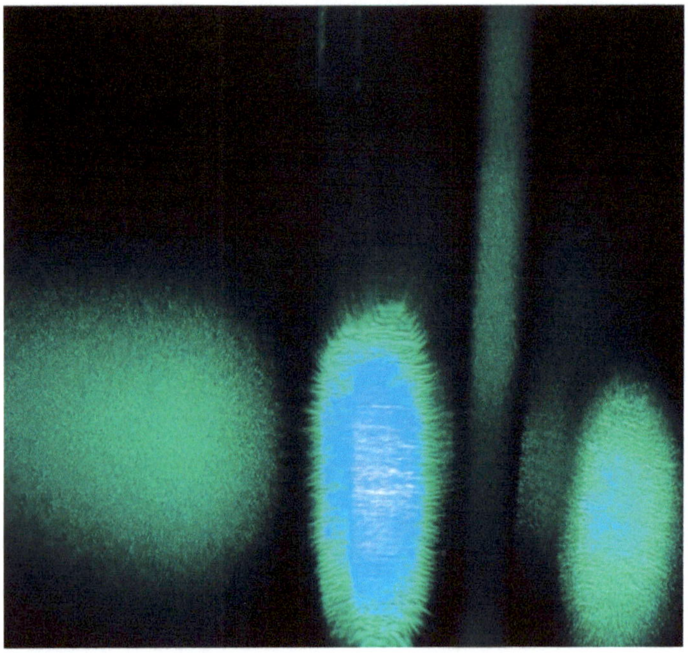

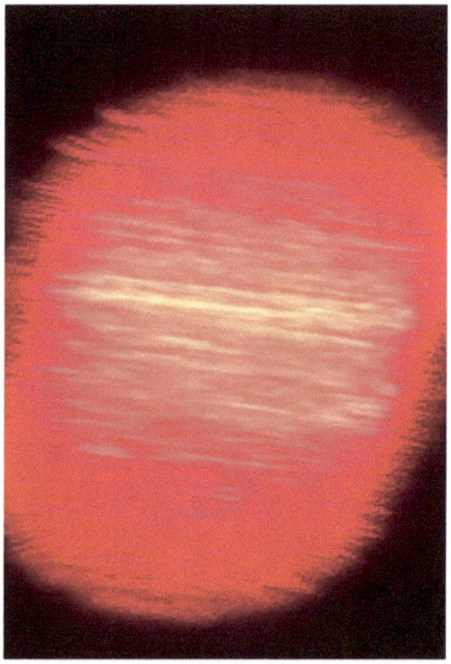

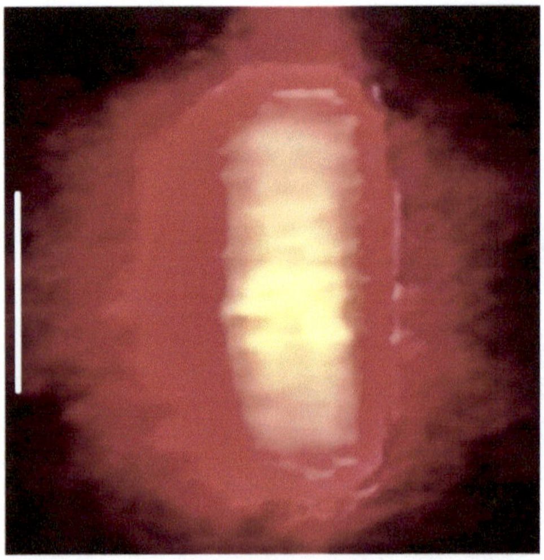

extrait from the video, we see in the mirror of the 0.07 nm wire, the image of light source(the laser emiting surface)

AND WHEN WE ZOOM IN we finally have our light source wich will be reflected on the 2-3-4...cylindrical mirrors represented by wires or slits

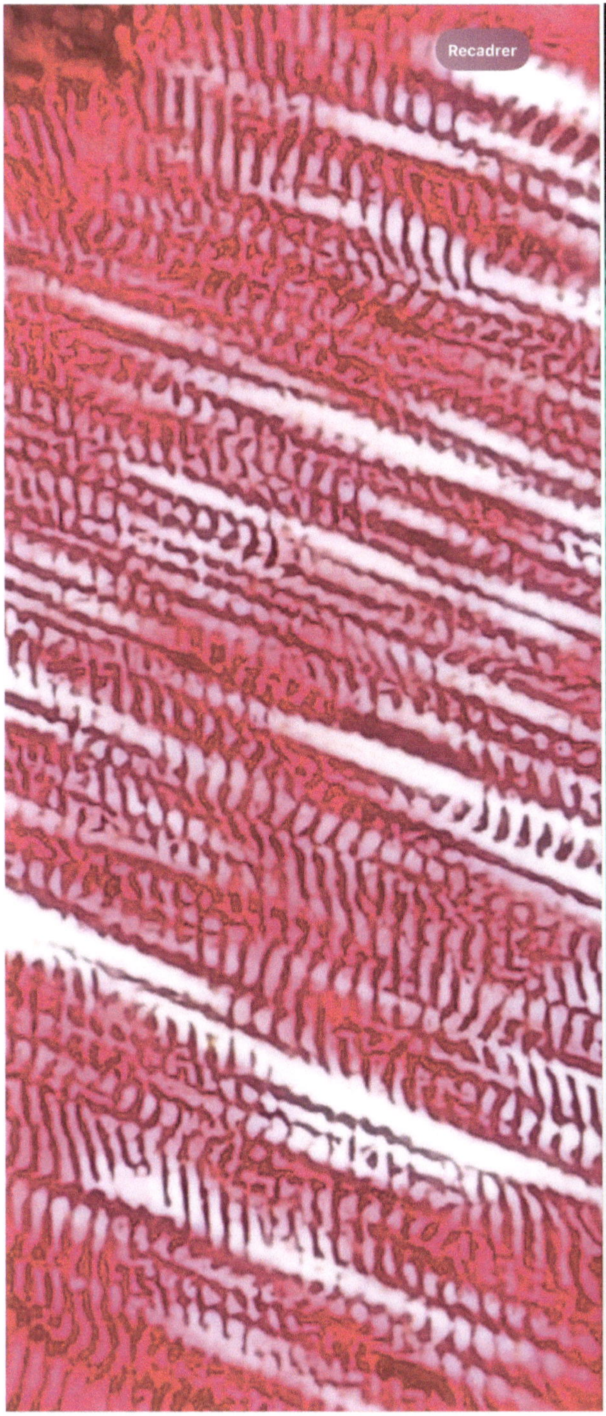

When we calculate what each filament reflected in the cylindrical mirror of the image represents, we count all the striations, and starting from the size of the screen on which we know we have the image of a wire like a hair, we obtain the red and green wavelenght

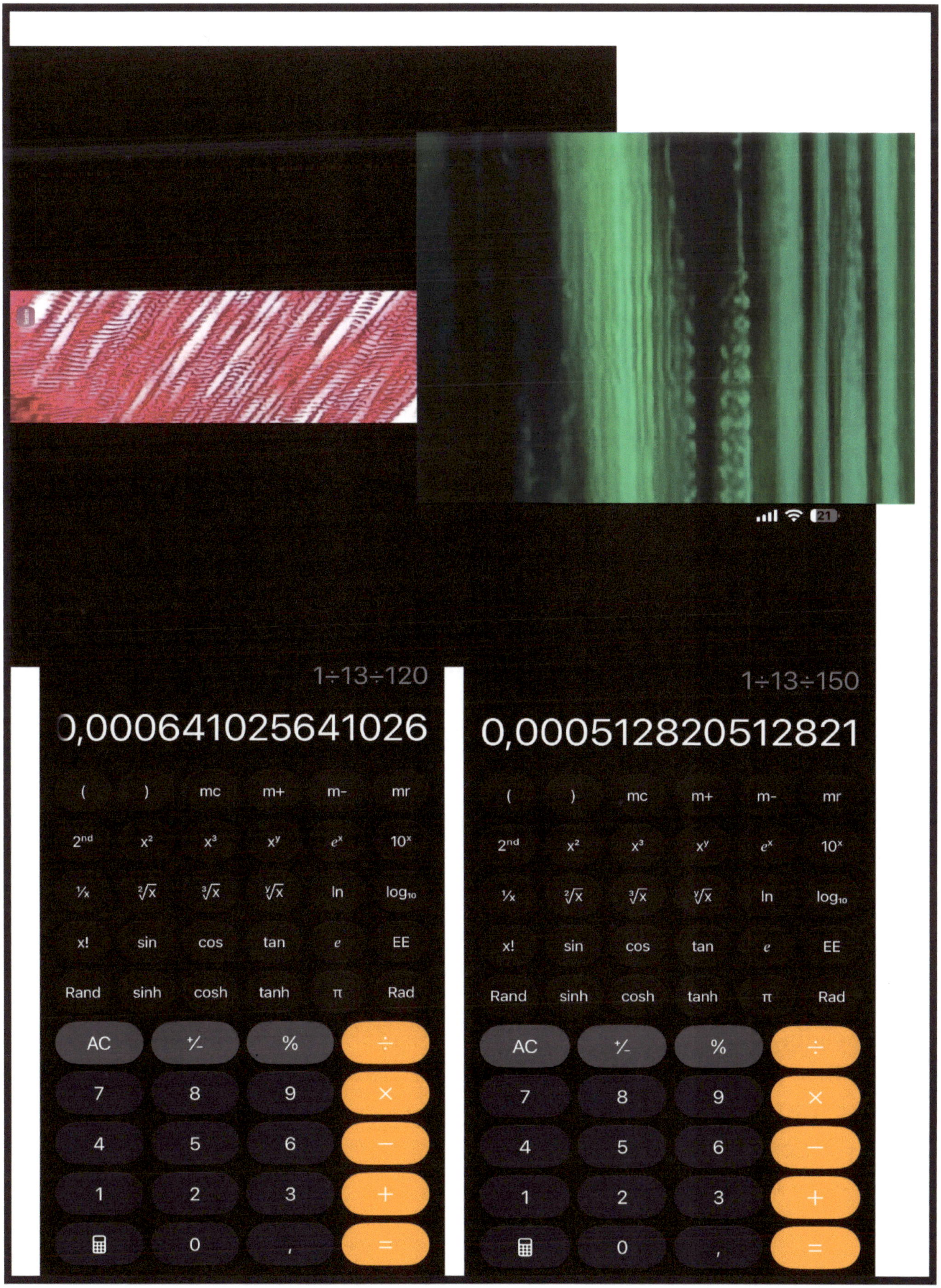
1÷13÷120
,000641025641026
1÷13÷150
0,000512820512821

Now comes the hardest part:
1.to understand how reflection on cylindrical mirrors occurs
2.how the reflection of different wavelenghts occurs at the atomic level and MUST IMPORTANT at the FRACTALONS level
3.What the frequency of light is ?
4.Why light and gravity have the same speed and what is the difference and the similarity between light and gravity.
Unfortunately I am not at all good as teacher or as a youtuber, so I will appeal enormously to you inteligence.I will present you visually the consequences to you and you try to understand them and assemble the puzzle.

Every image created by cylindrical or spherical mirrors(and practically everything we see with our eyes)can be decomposed into:
1.1 THE SHADOWS created by two mirrors, in our case the two wires or the the space between the 2 slits(2 cylindrical half mirrors and a central cylindrical mirror)
1.2 The paths of direct light and the paths of reflected light in front and in the back of mirrors
1.3DIRECT LIGHT from the source,the importance and the pattern of differentes wavelenghts, how that light is created and how in every source of light we already have a direct light and a reflected light ,reflected in the superficial layers of the vacuum upon contact with source after its creation and will give us a different pattern of the light source for each wavelenght, like a kind of ocean of standing waves at resonance, a bucket of water with square walls that is hit at a certain frequency by the bottom.
1.3 REFLECTED LIGHT AT THE SURFACE OF CYLINDRICAL MIRRORS depending of the wavelenght and creation of the base of DSE pattern, the smallest pixel ,the lenghtwave POINT on the wall creating all the image

WE will start with the easiest part
1.1 THE SHADOWS

1.1 The shadows created by two mirrors, in our case, the two wires or the two slits(two cylindrical half mirrors and a central cylindrical mirror)This give us the first pattern in DSE that will repeat, centrally direct light and shadow

AND ATOMIC-LEVEL SIMULATION OF REFLECTION SURFACES

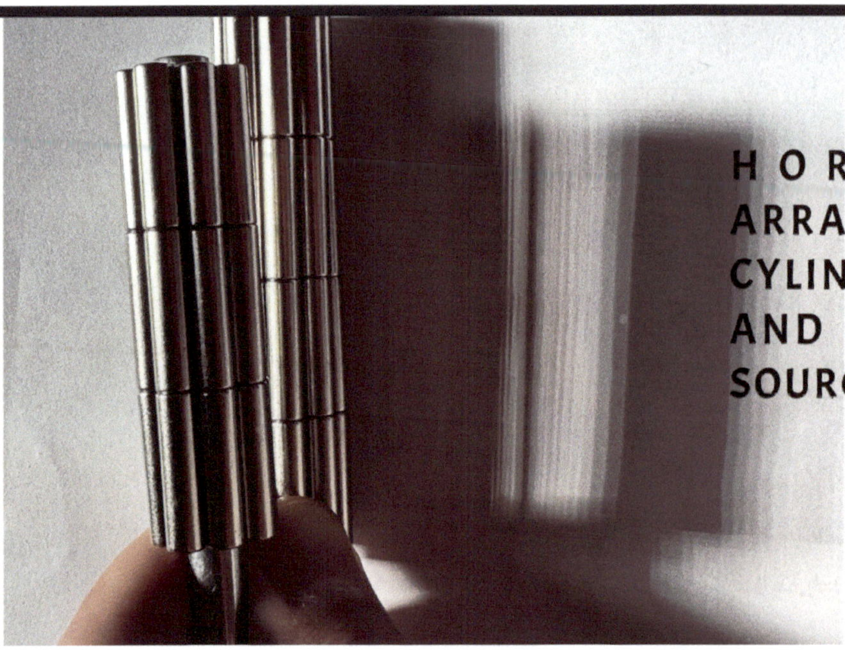

HORIZONTALLY ARRANGED PARALLEL CYLINDRICAL MIRRORS AND VERTICAL LIGHT SOURCE

AND THE MOST IMPORTANT IMAGE TO UNDERSTAND DSE is TWISTED CYLINDRICAL MIRRORS

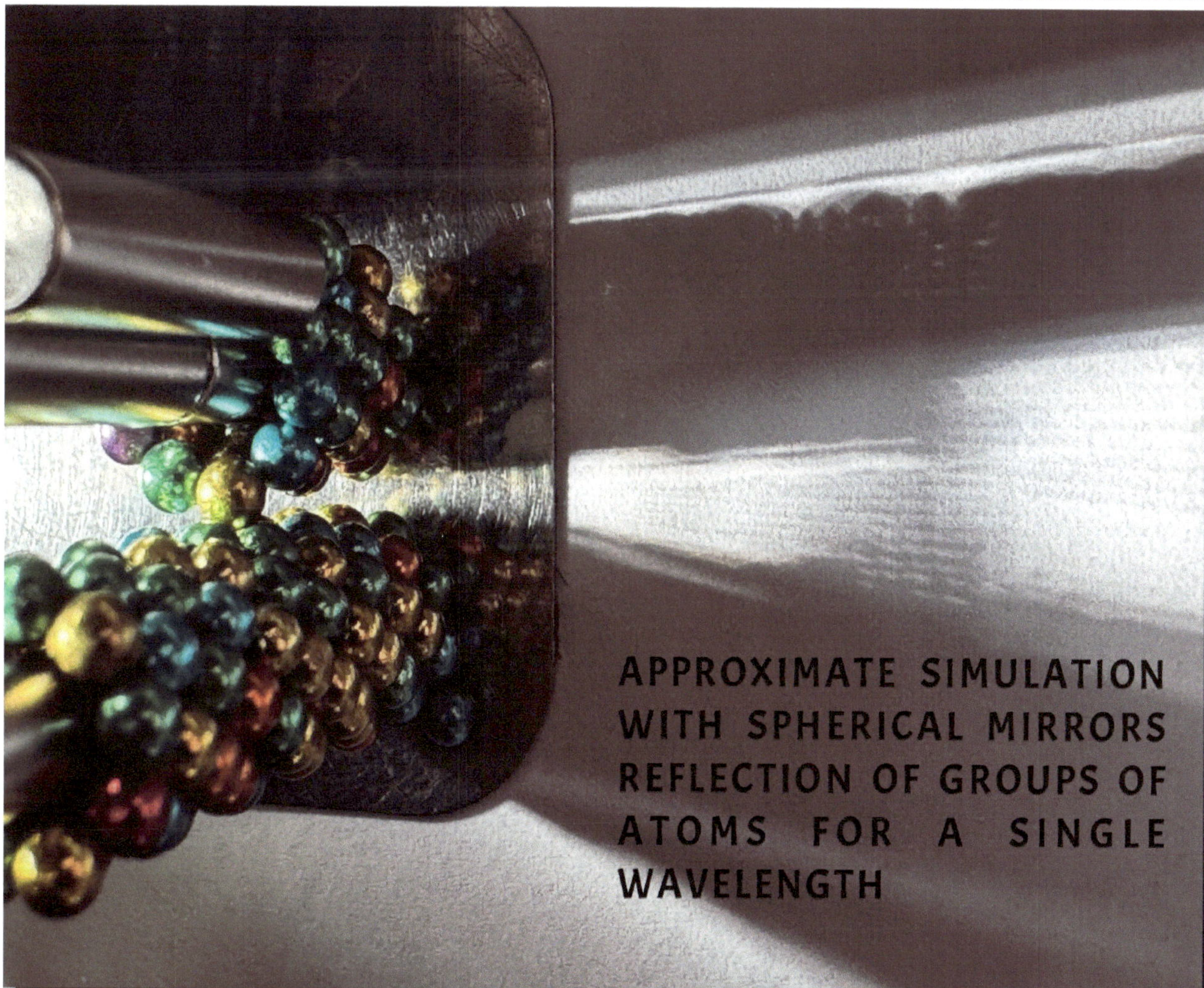

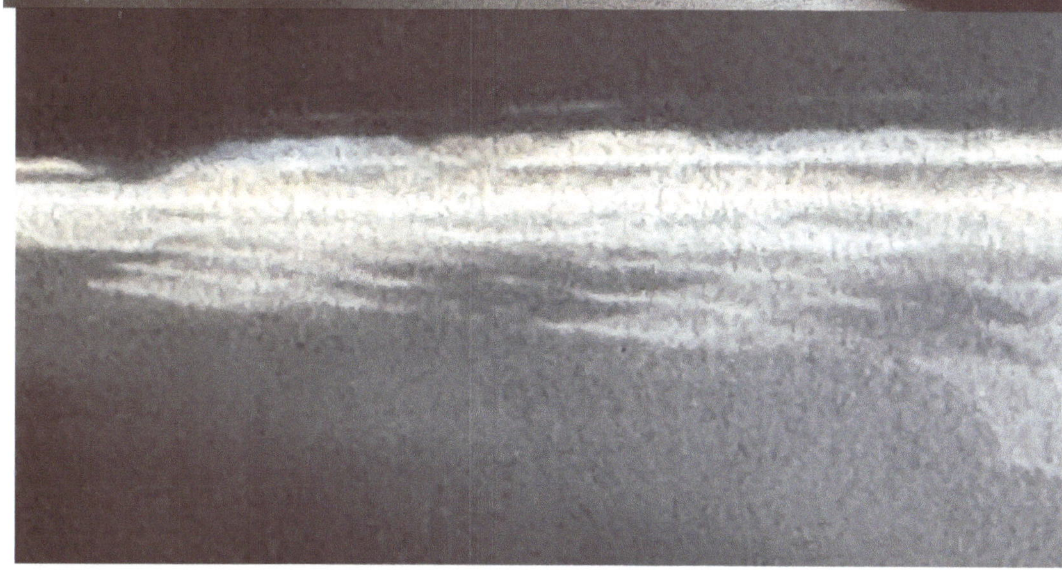

Zoom of the smallest elements in the image, the base of the entire image of shadow and light

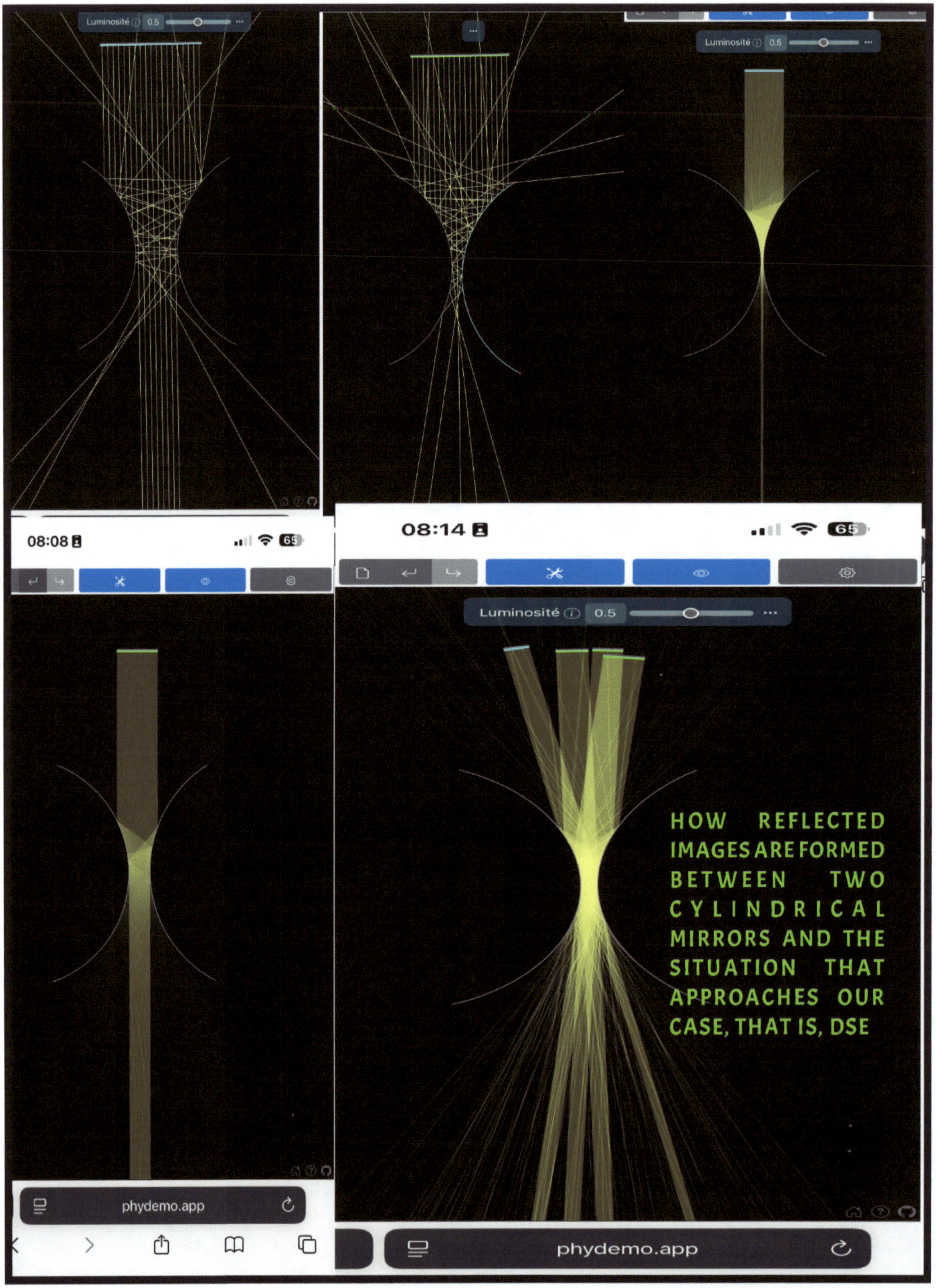
Luminosité 0.5
08:08
08:14
Luminosité 0.5
HOW REFLECTED IMAGES ARE FORMED BETWEEN TWO CYLINDRICAL MIRRORS AND THE SITUATION THAT APPROACHES OUR CASE, THAT IS, DSE
phydemo.app
phydemo.app

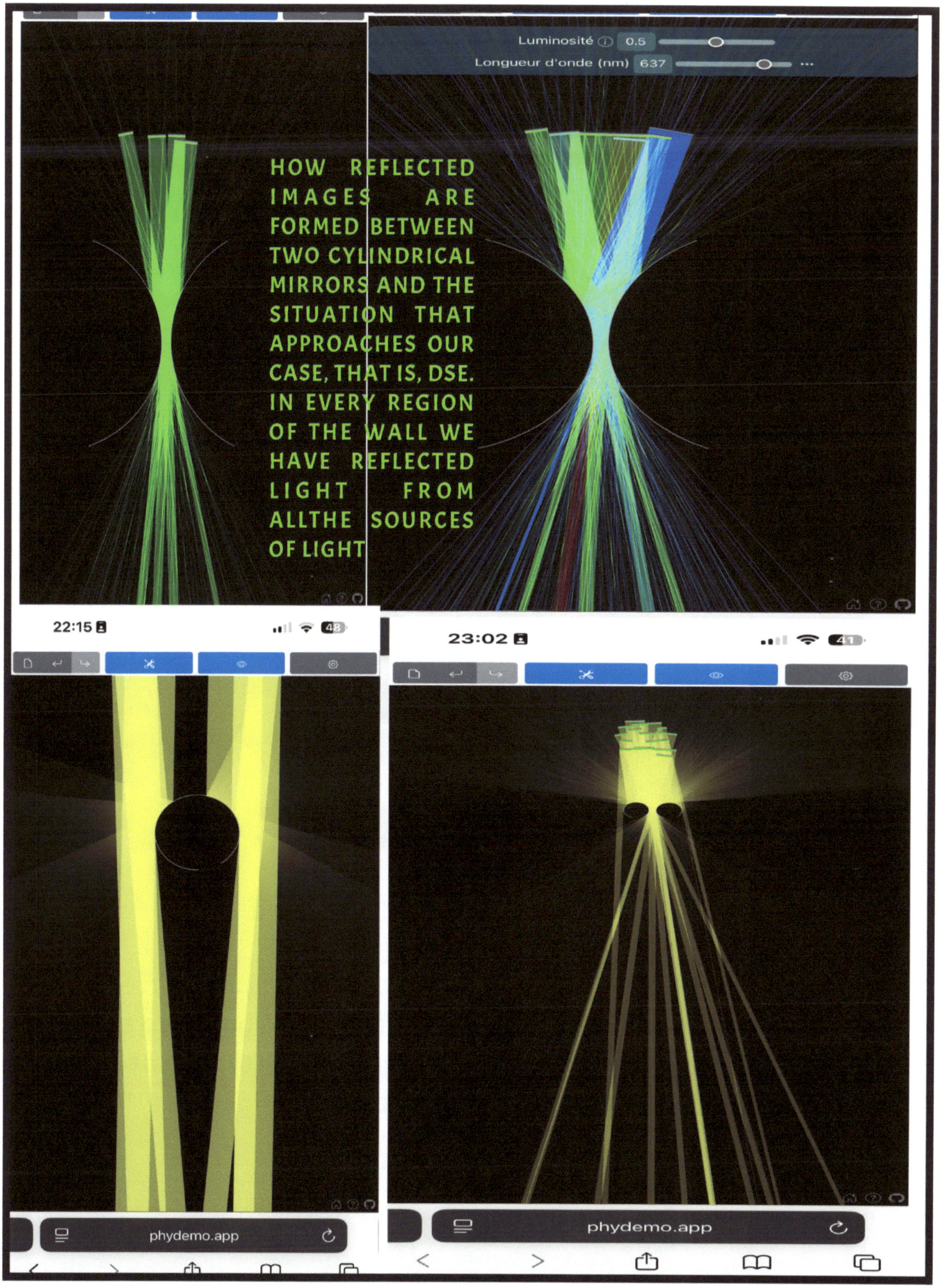

HOW REFLECTED IMAGES ARE FORMED BETWEEN TWO CYLINDRICAL MIRRORS AND THE SITUATION THAT APPROACHES OUR CASE, THAT IS, DSE. IN EVERY REGION OF THE WALL WE HAVE REFLECTED LIGHT FROM ALLTHE SOURCES OF LIGHT

THE SHADOW PATTERN, at different zoom levels and different shooting angles, the pattern changes completely, but if we zoom out on the photo we will also find the one in the other photos, it is already coded in the colors in the photo

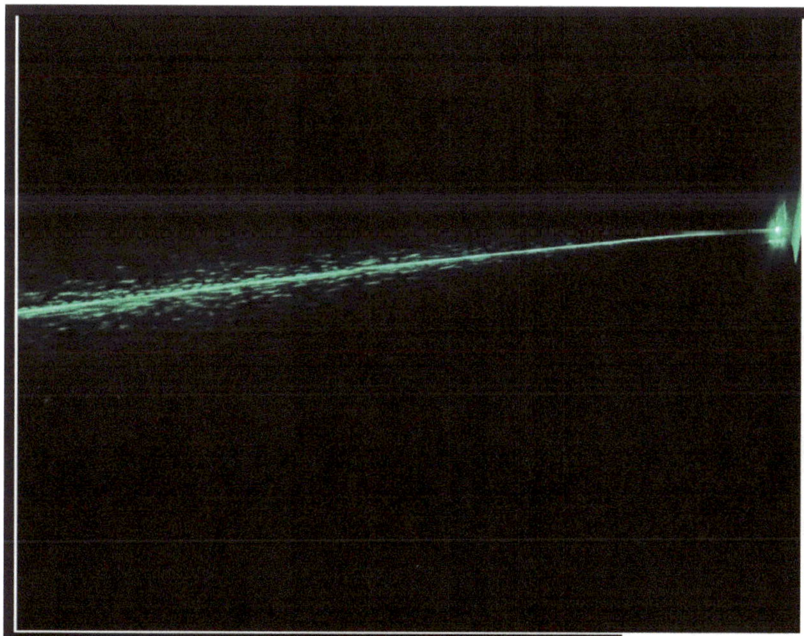

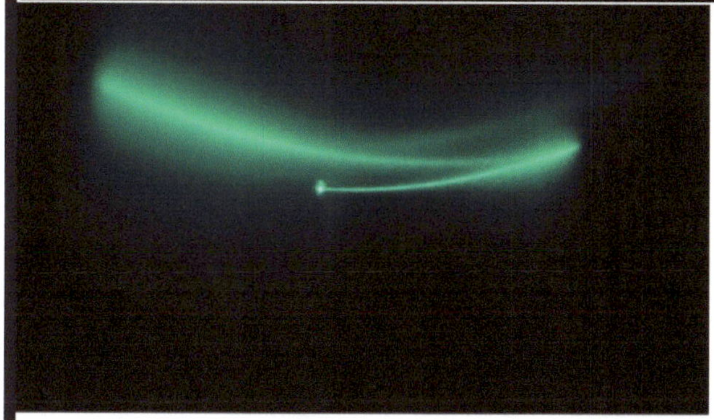

If we tilt(vertically and horizontaly) the laser between the two cylindrical mirrors, we separate and we obtain the second element that will be reflected in final image, the directly reflected light, which will be repeated for the lateral observers on the left and right

We can do it the DSE with light ,electrons or footballs the reflexion is the same

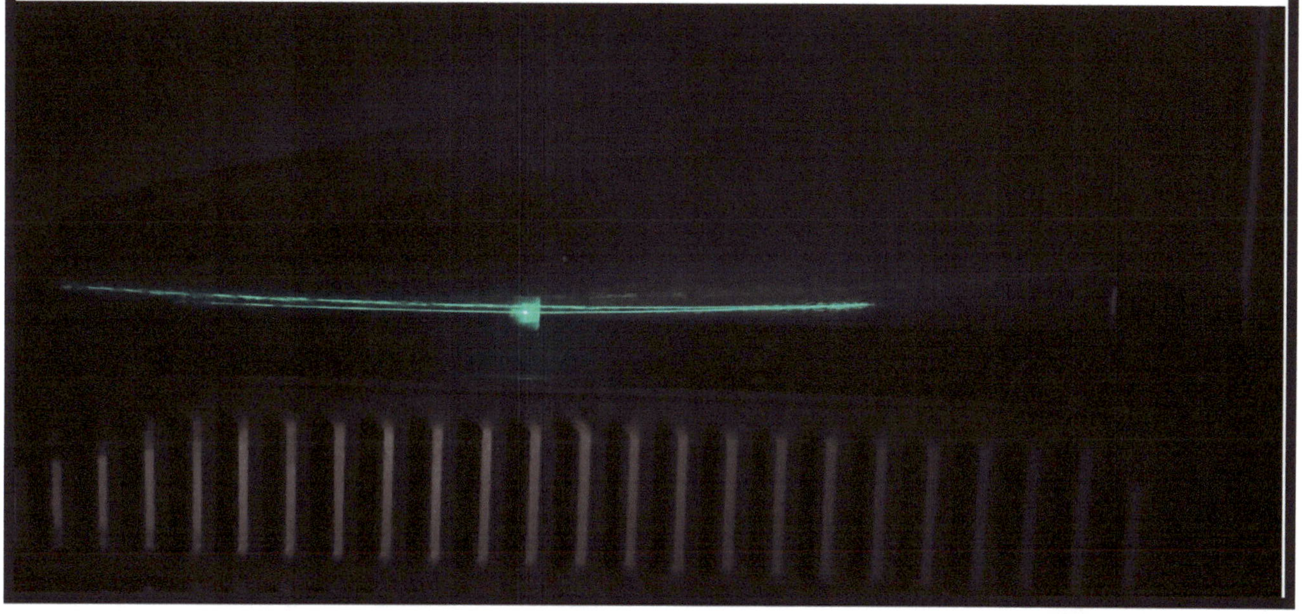

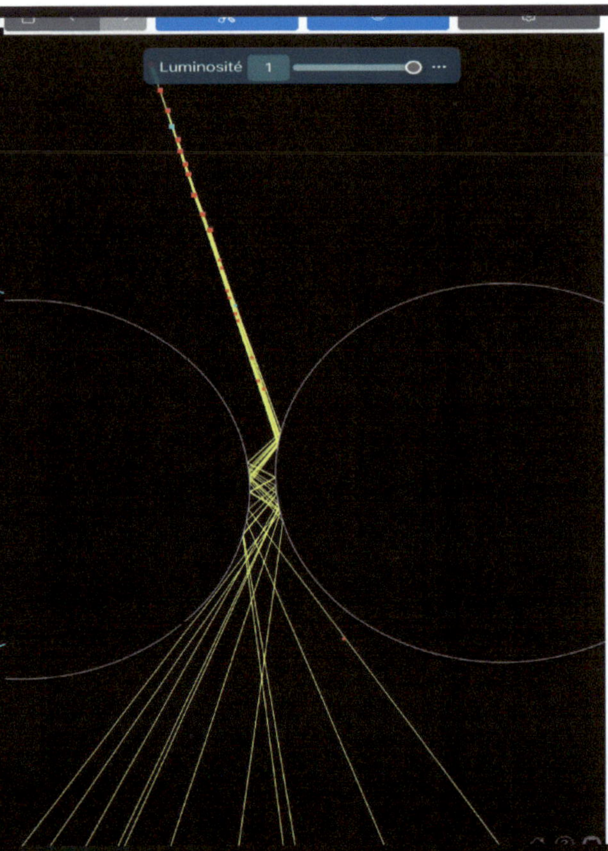

Luminosité 1

SG

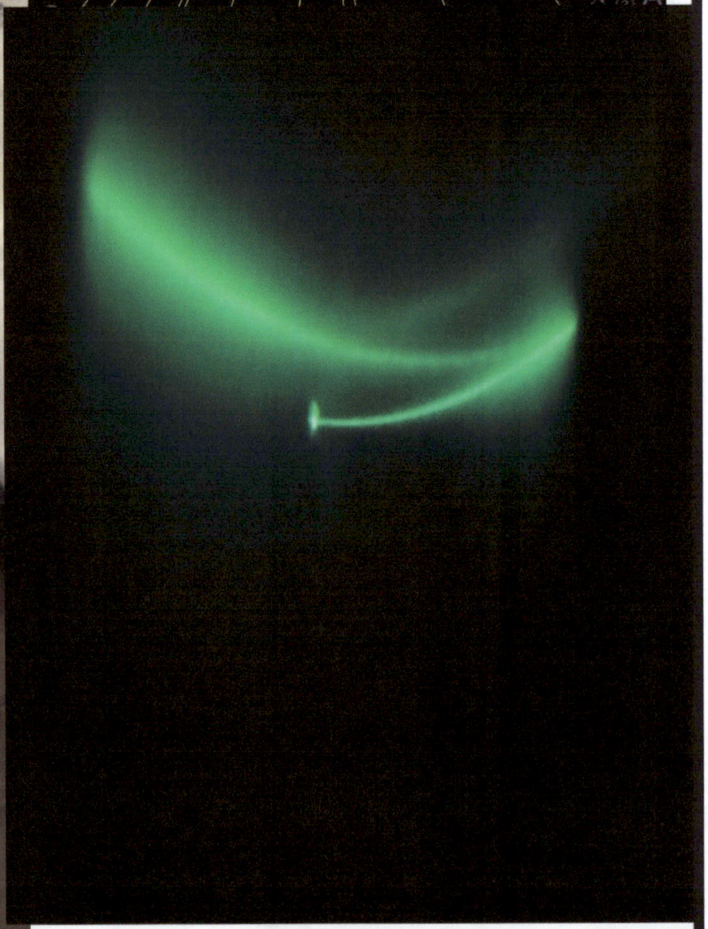

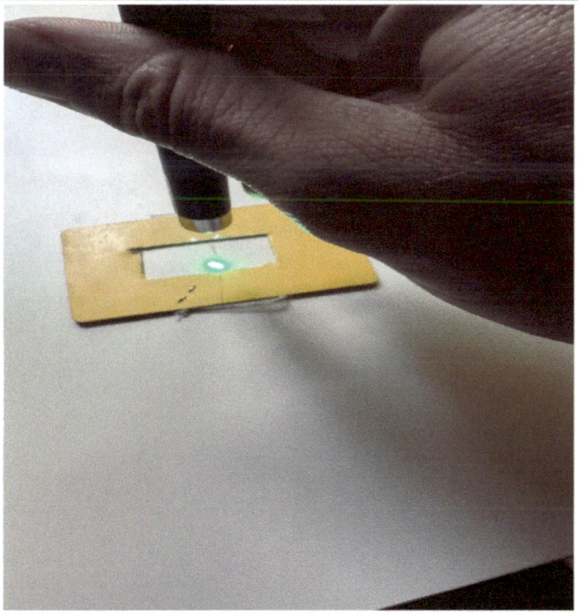

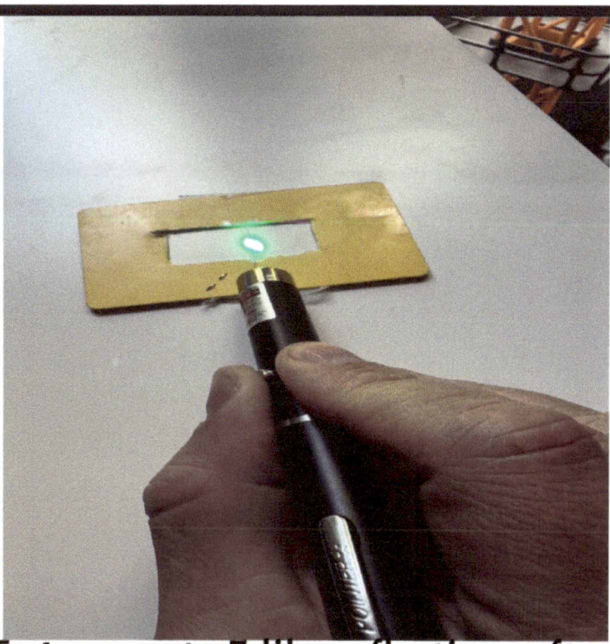

And the same phenomenon in DSE, 4 separate Z like reflections after 1.2.3.4..reflections, we decomposed the linear image into its large components, in the next part we will deal with the essential part, the "deep pixels" i.e. the wavelengths from the light source.Most people don't know that the image obtained in DSE is always perpendicular to the axis of the slits, and when we decompose it the 4 images are and move if we hold the slit twisted in our hand, it moves at 45 degrees to the reflection surface of the 4 cylindrical mirrors each independently of the others (video on fractalon.fr)

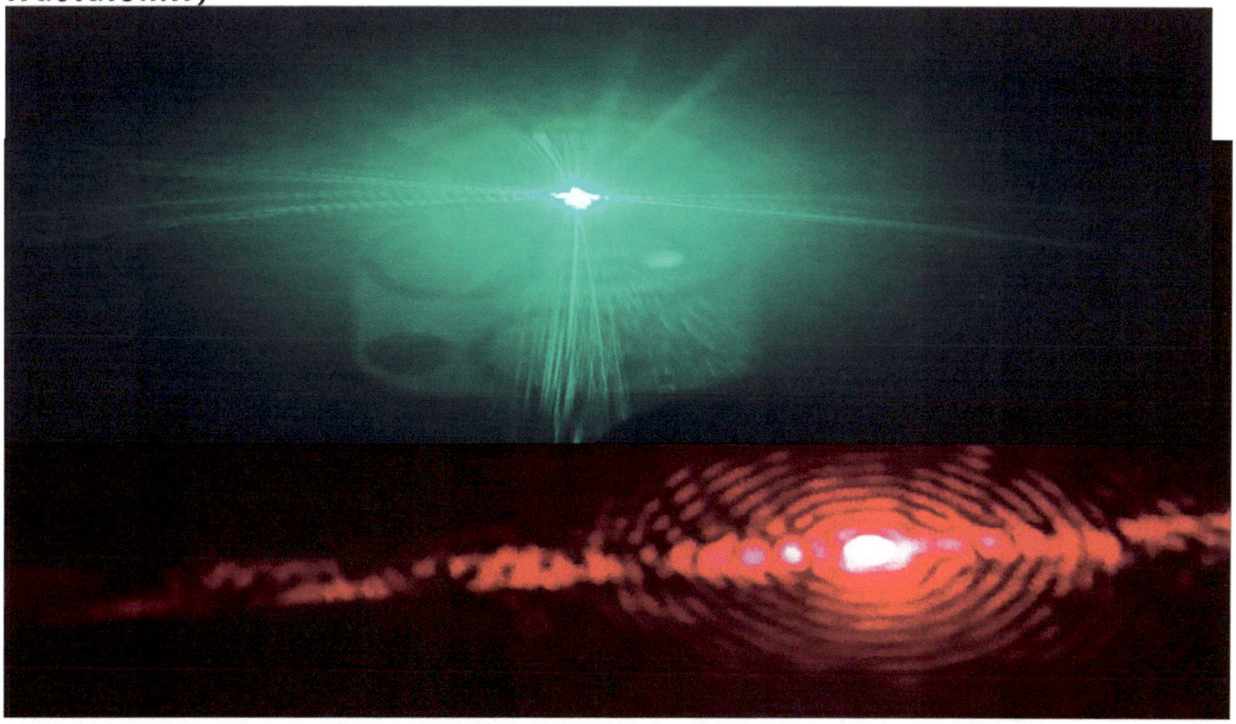

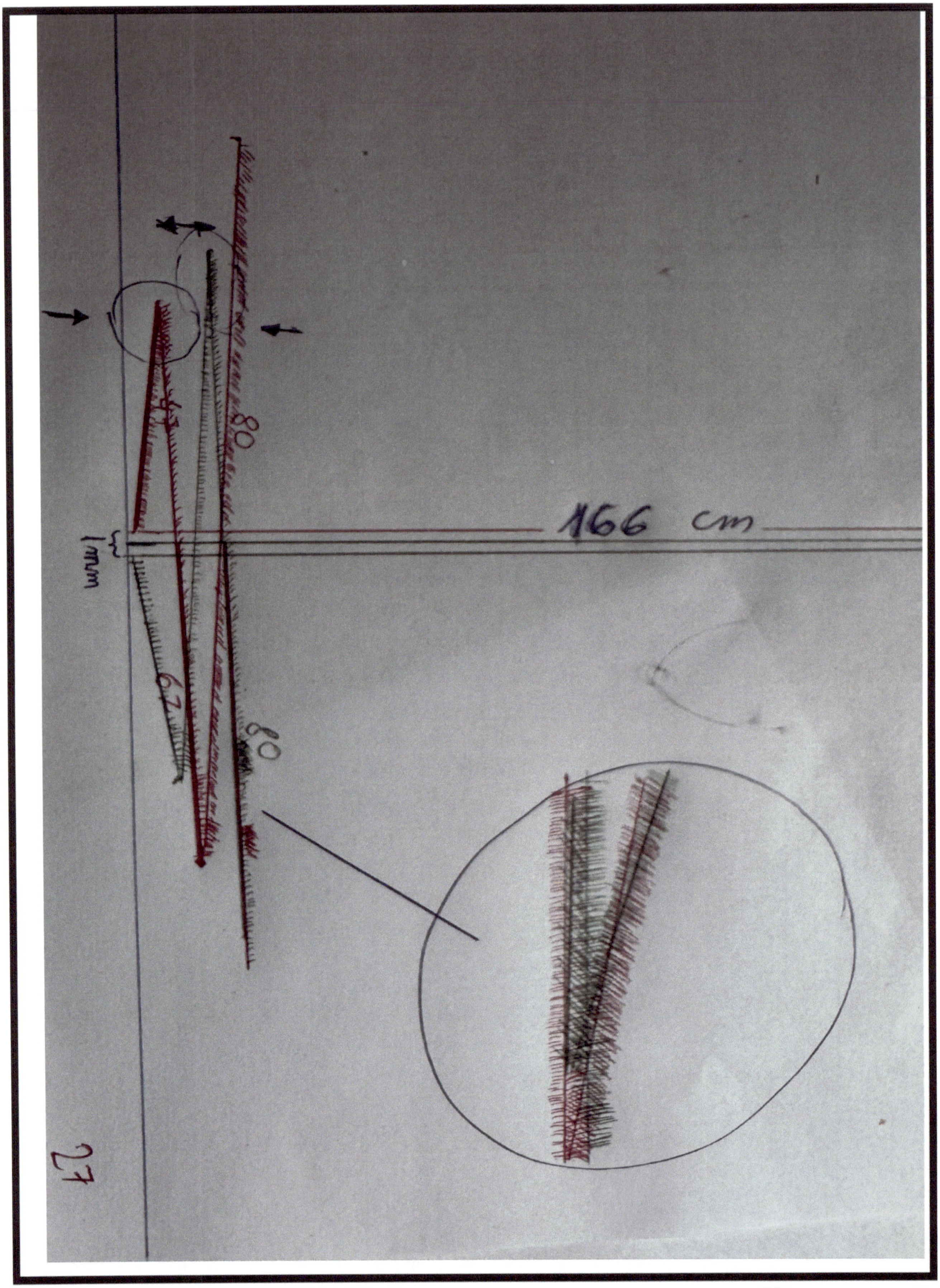
166 cm
80
67
80
67
1mm
27

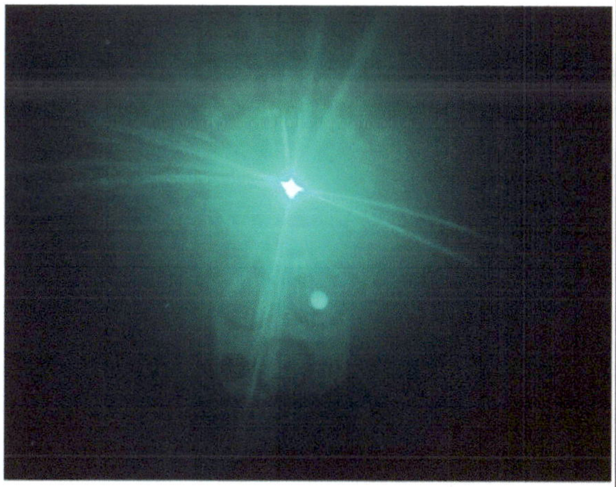

inclined wires,separeted Z reflections

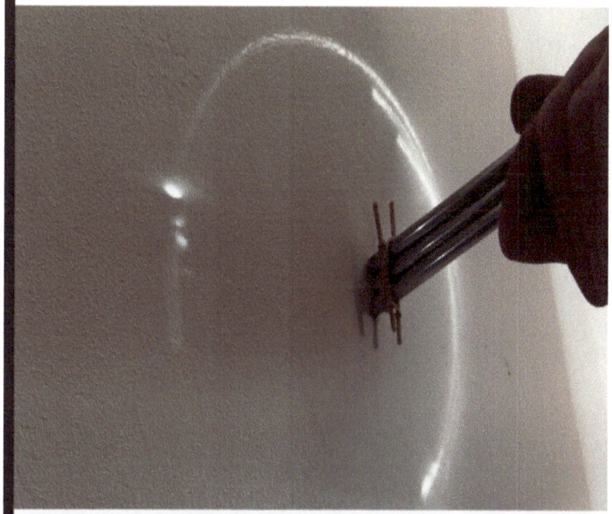

In the same image we have the DSE of light reflected in front of the 3 inclined forks with the front side covered by non reflective surface (cylindrical mirrors) and behind them. They will always form a perfect circle if they are perfectly parallel. so what we call refraction today is still a reflection at the atomic level, I will explain it to you

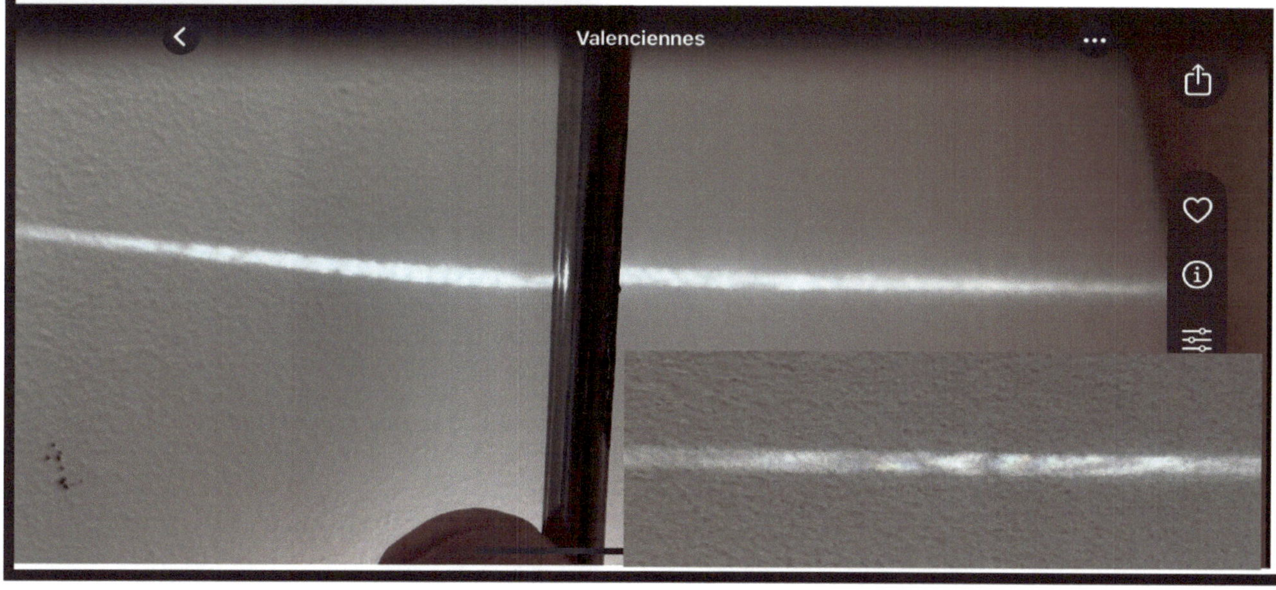

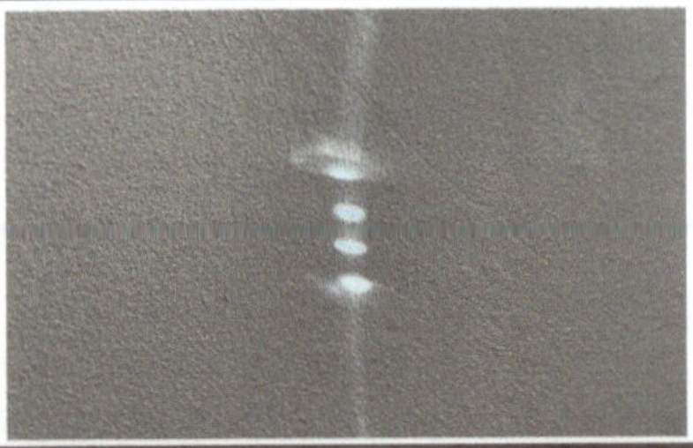

DSE examples with reflected image in front and behind the mirrors

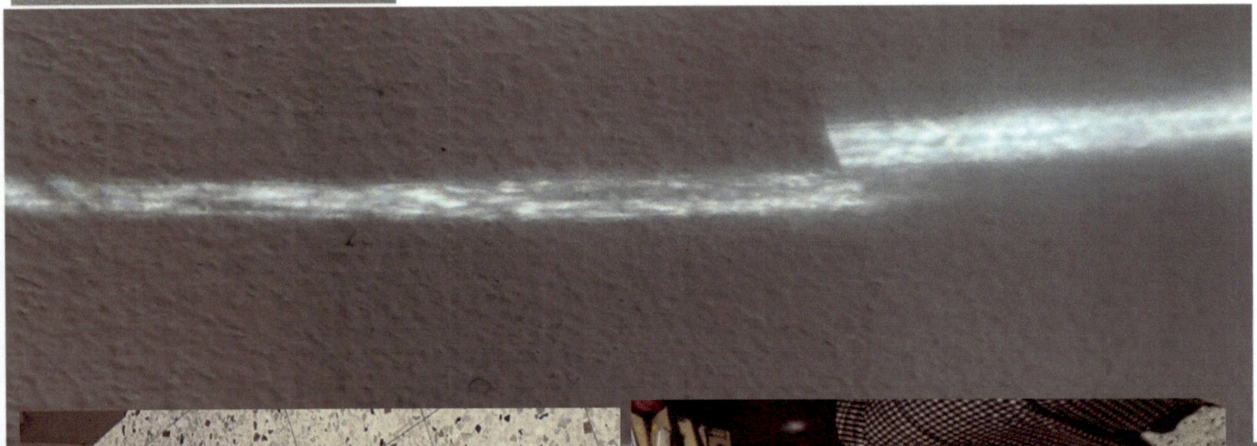

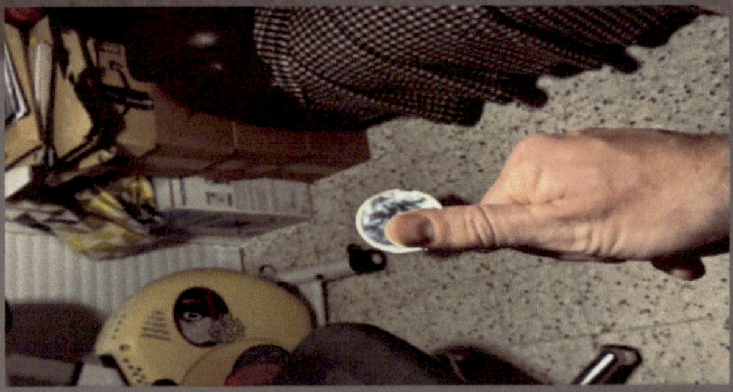

DSE only with reflected light of sun in the front of the mirrors, you must choose a right angle of left or right observer for the 3 cylindricals mirrors,at nanoscale is impossible to block the direct light from the image like in the two photos in the center

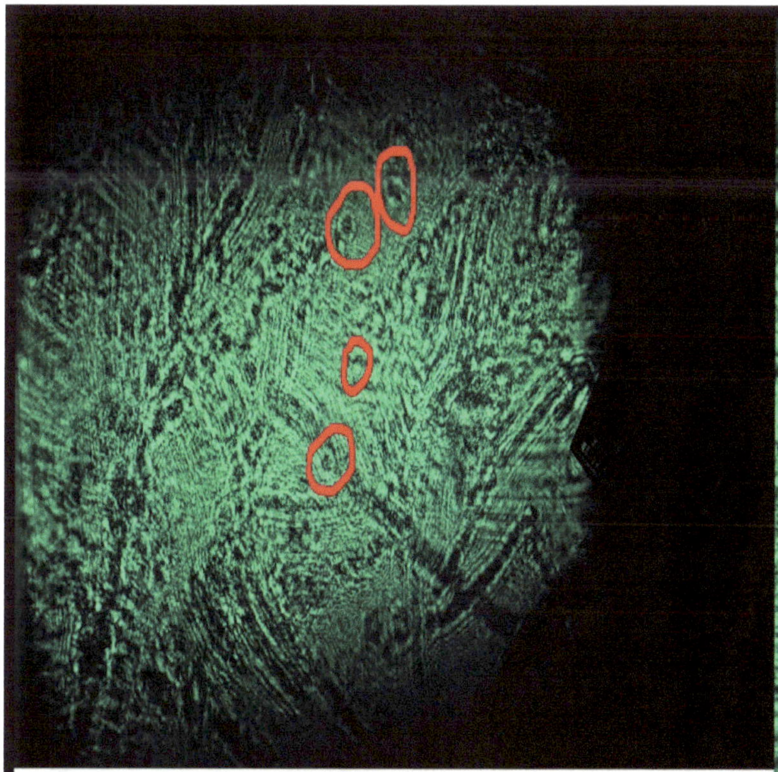

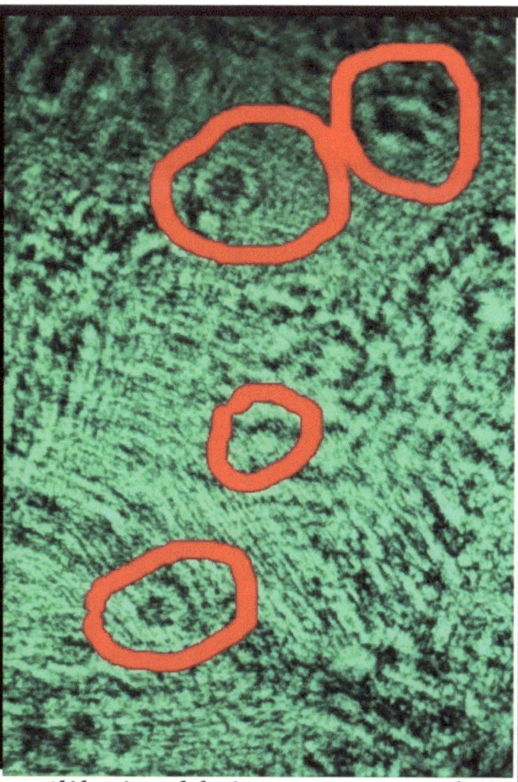

When you send a light source onto a lens surface (like in old cinemas or modern projectors), the smallest dust particles on the lens, or colors in the image, will always create concentric DSEs, and the sum of these images will create the image on the screen. If we focus on the focal length that gives us all the parallel rays of direct light and not the reflected ones, we get a clear image on the screen.

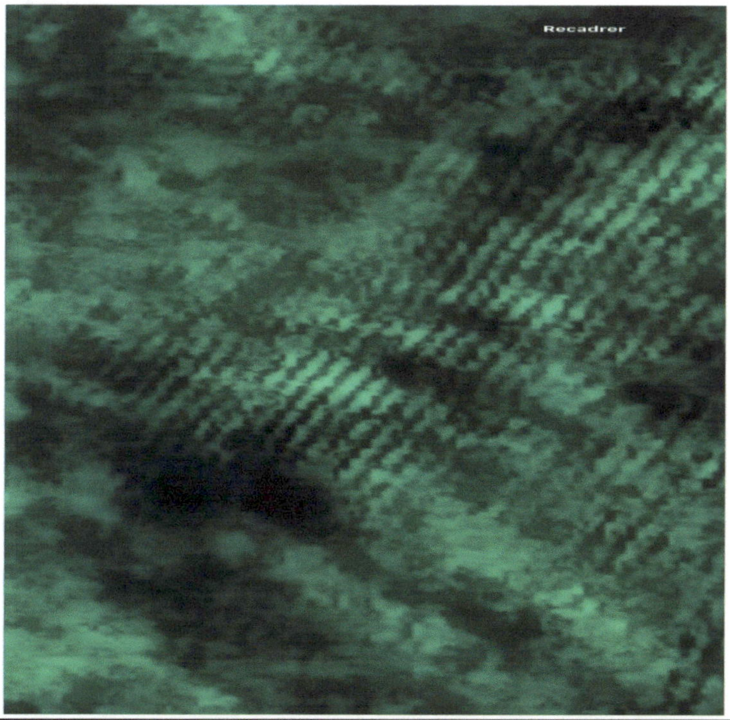

but in each image for a given color the smallest visible element depends on the wavelength of the light from the source and the pattern in the source

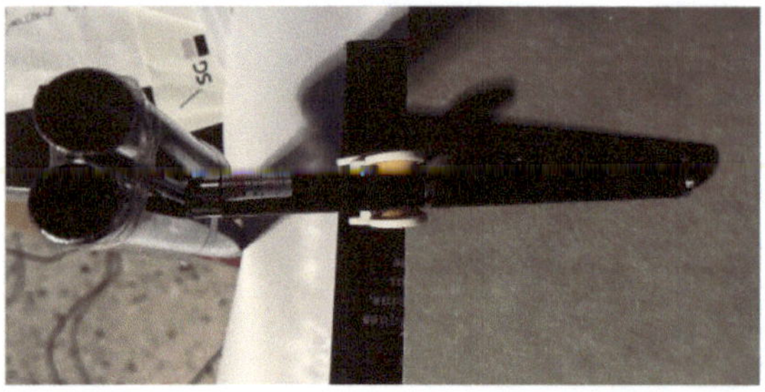

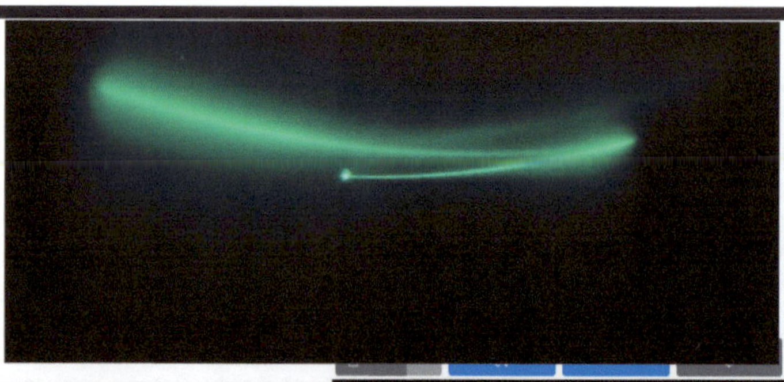

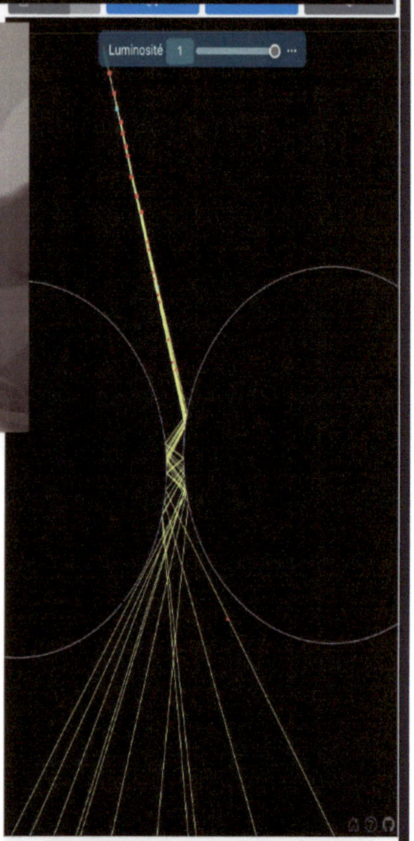

For different wavelengths and for different distances between the walls(mirrors) of the slits in the DSE we can measure how much the corners of "z" move laterally for each point source of light coming from the laser(don't forget our projector is giant, the two images at the beginning of book obtained by photographing the wires of the DSE)

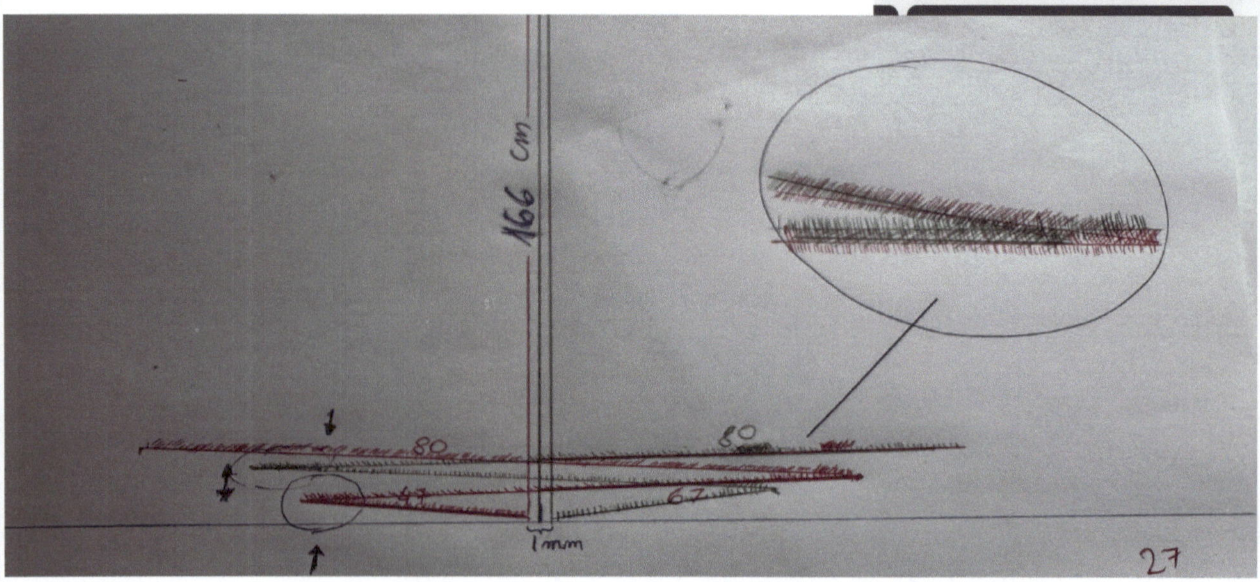

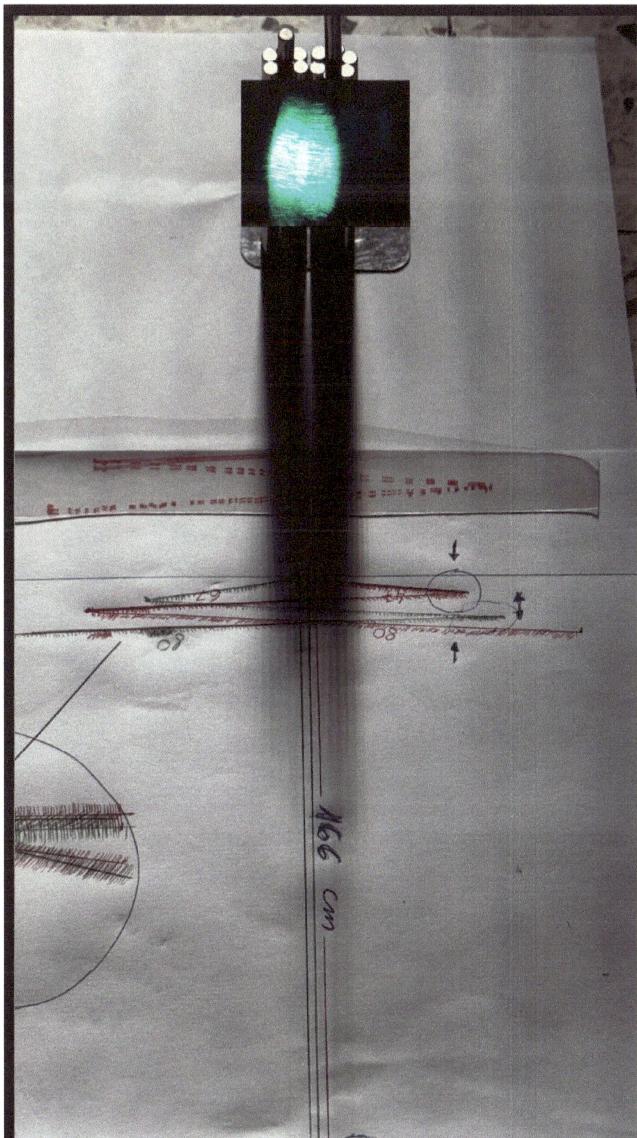

To imagine how each quantum(pixel) of red or green light will position itself, somewhere on a precise position on the "z" of reflection that corresponds to it, exactly offset by the wavelength from its neighbor in the light source

We will have a sequence of multiple intersections between parallel reflected lights, at distances corresponding to the wavelength. At different distances we will obtain different patterns of bright spots with light reflected once, twice, etc. from all the nearby bright spots in the laser.

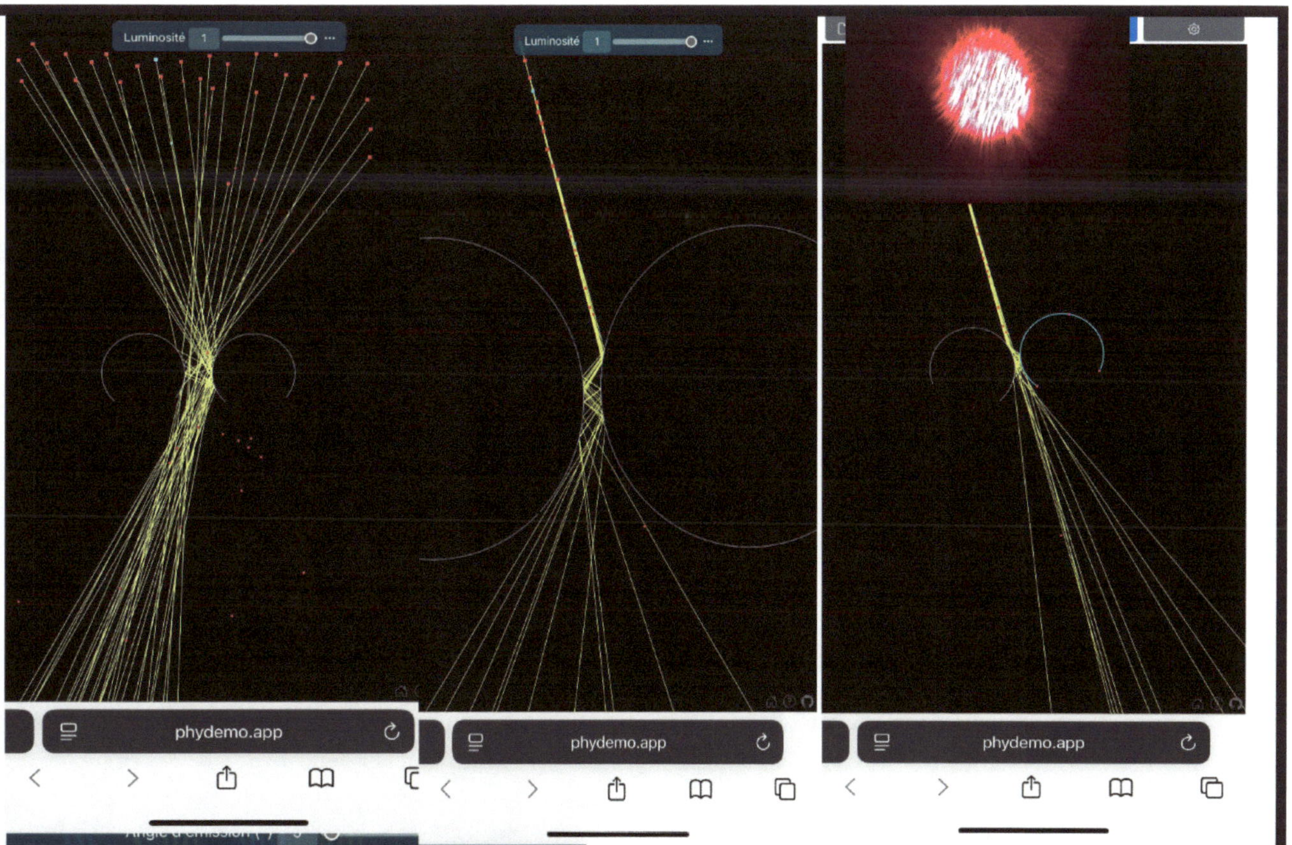

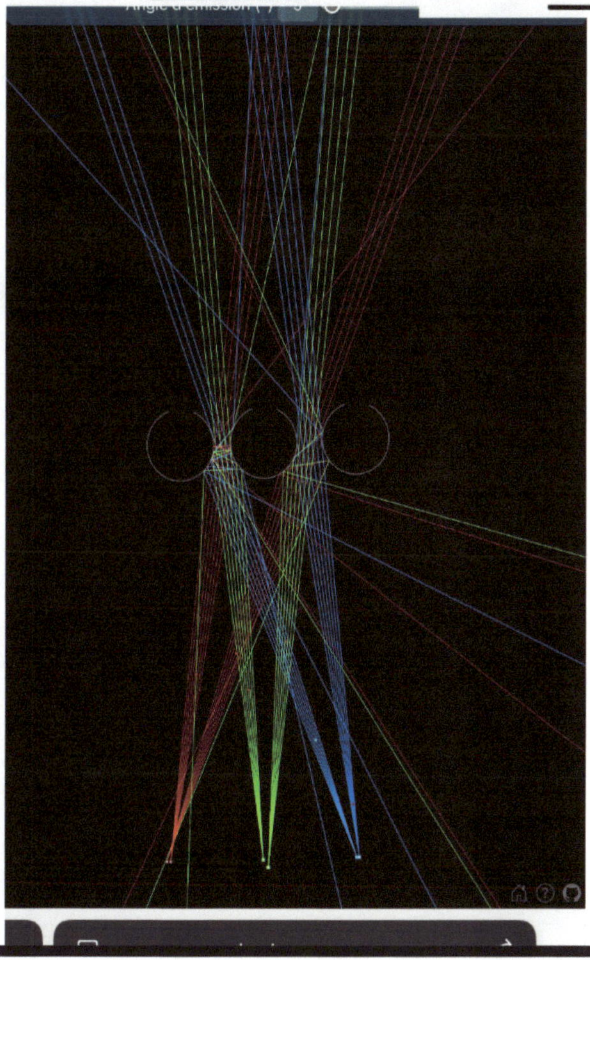

Bright points on the same line and the same column will be found in different parts of the DSE

To understand DSE, the easiest way is to look backwards from the observer's point of view, at what we see reflected in the mirror in the hedgehog image from the laser.

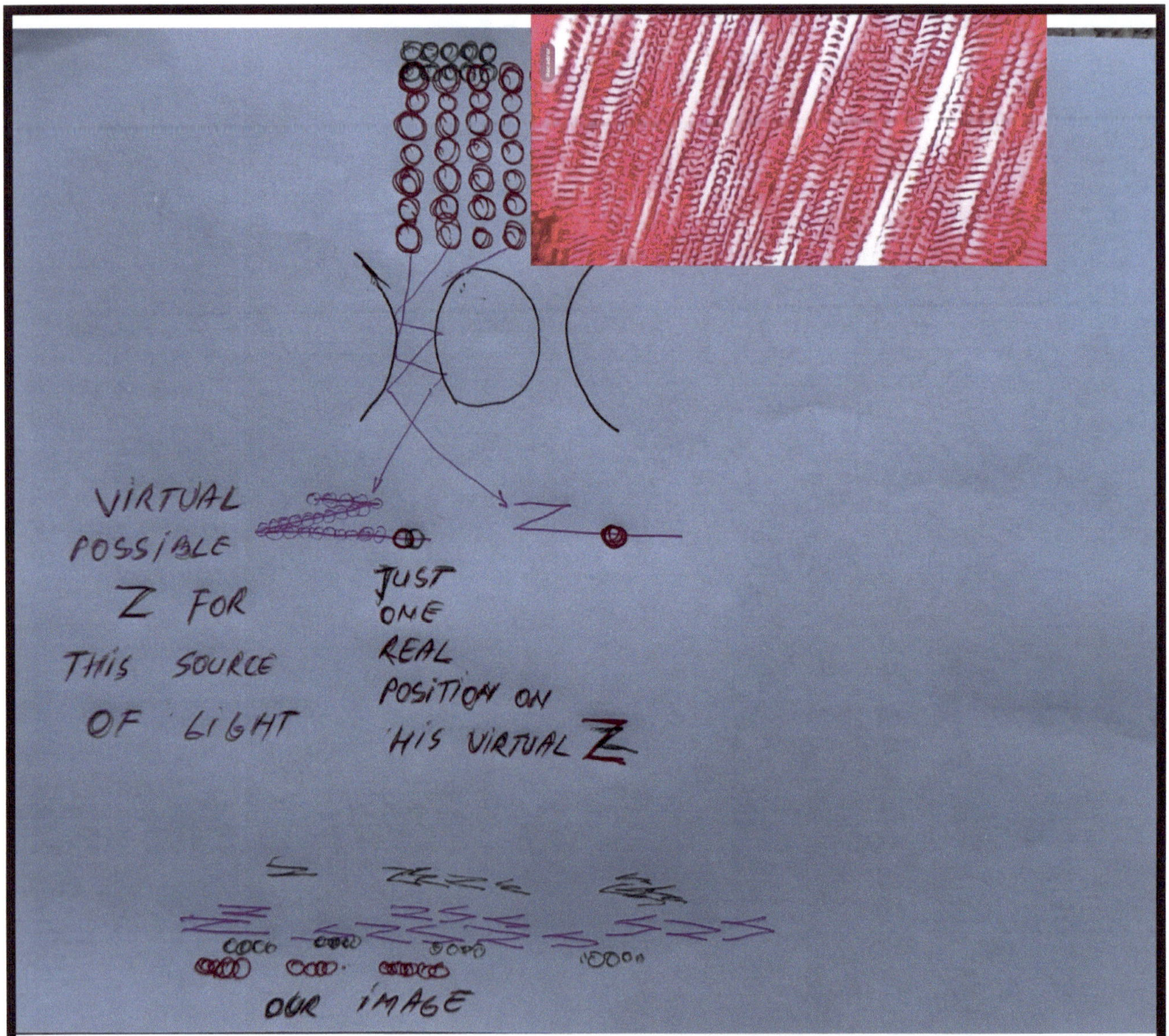

The final image is obtained from the superposition of all virtual "z's" on which there is a single real bright point corresponding to each bright point in the laser

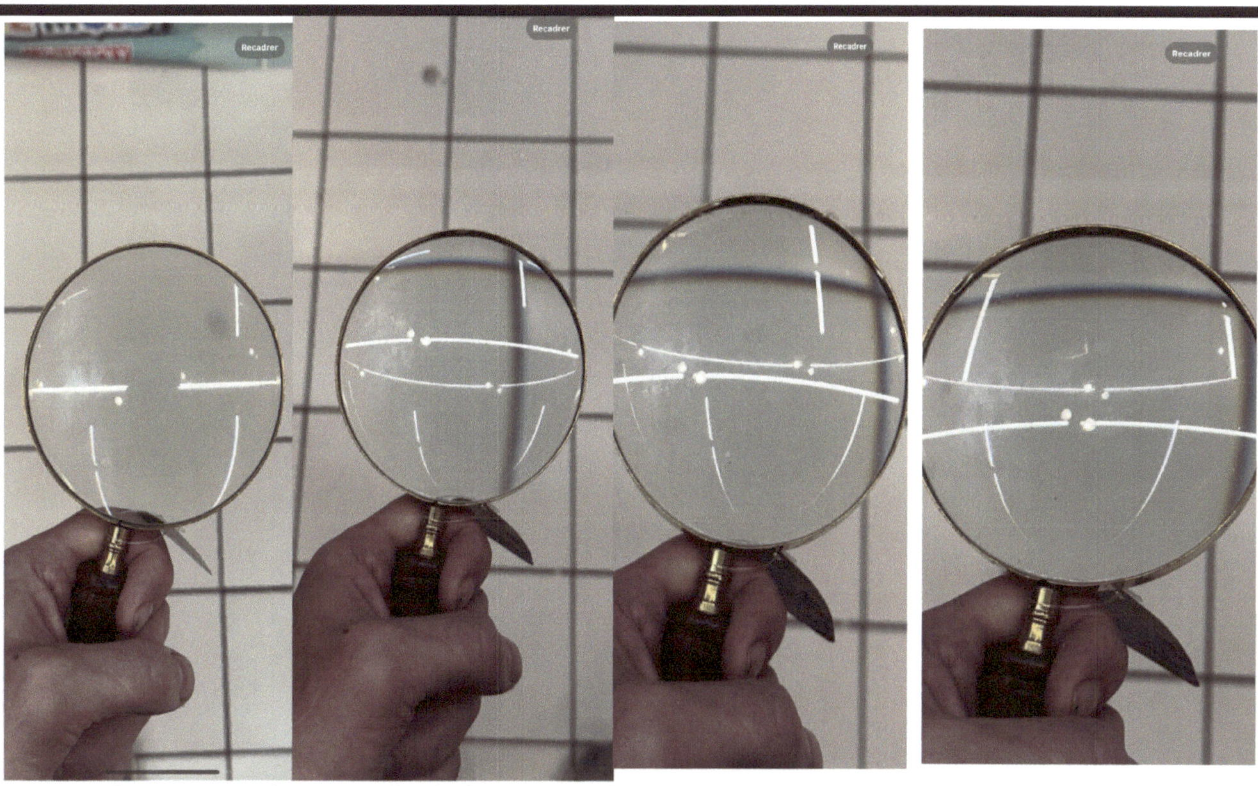

How the image of a straight light line is formed in spherical or cylindrical mirrors. It will always form a circle inclined at a certain angle to the real plane of the light source. The image in the DSE corresponds to the first image with the clarification that we do not have only two reflections we have a multitude "z". By tilting the slits, the wires and the wall we can recover the circles of which all similar reflections are part (to obtain the image I stood with my head and the hairs in front of the slits. The central vertical line represents the image given by the DSE on the untilted copper wires.

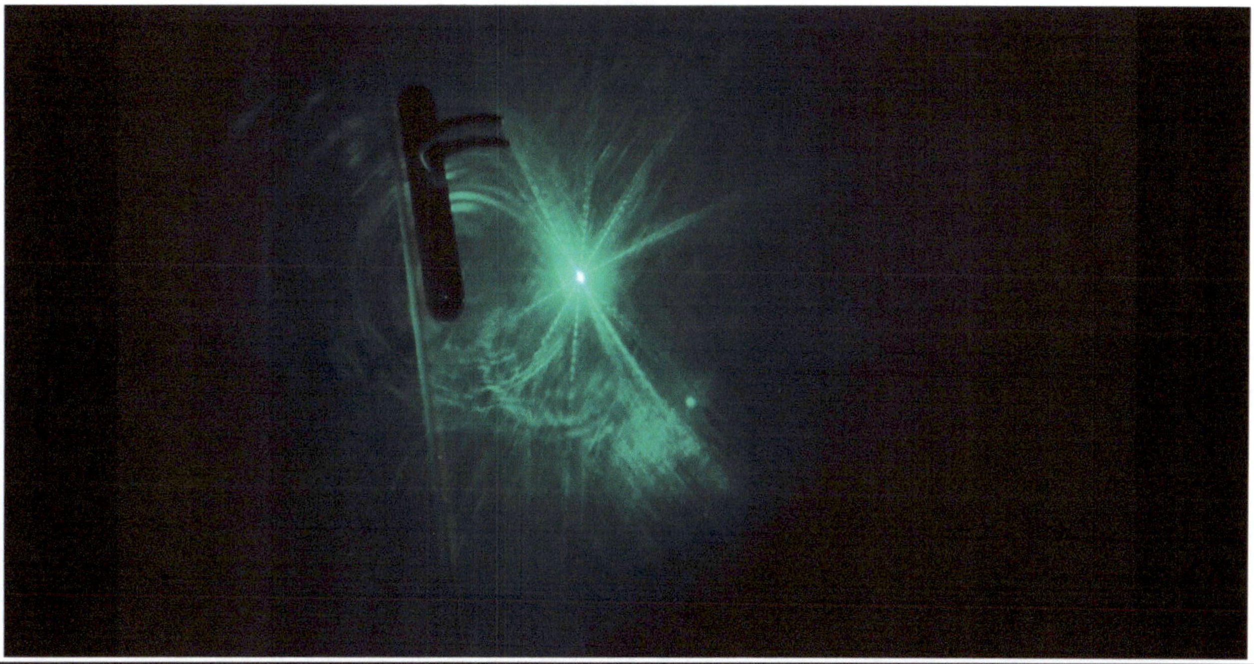

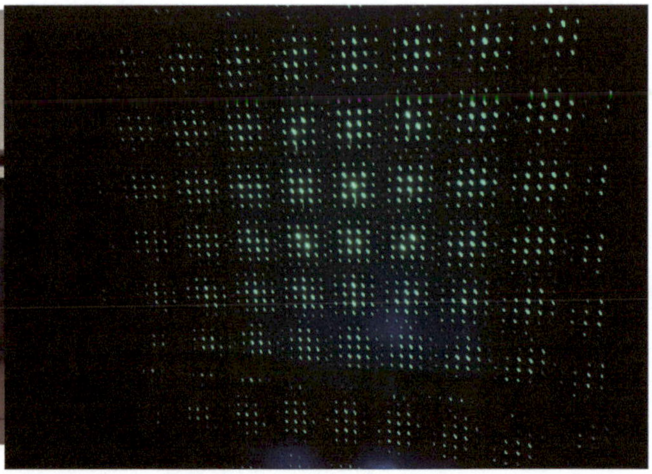

To better understand what is happening and to more easily decompose the image, we will do a simulation on a larger scale between the mirrors, on a single mirror, and the z-s on both mirrors.

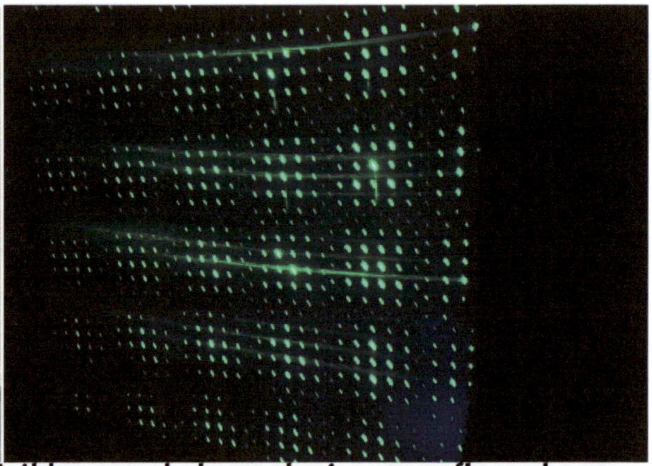

From all the points that are no longer visible we only have the image reflected once, twice, three times, with the distance between points equivalent to the distance between the light sources, each z is offset by this distance from the next

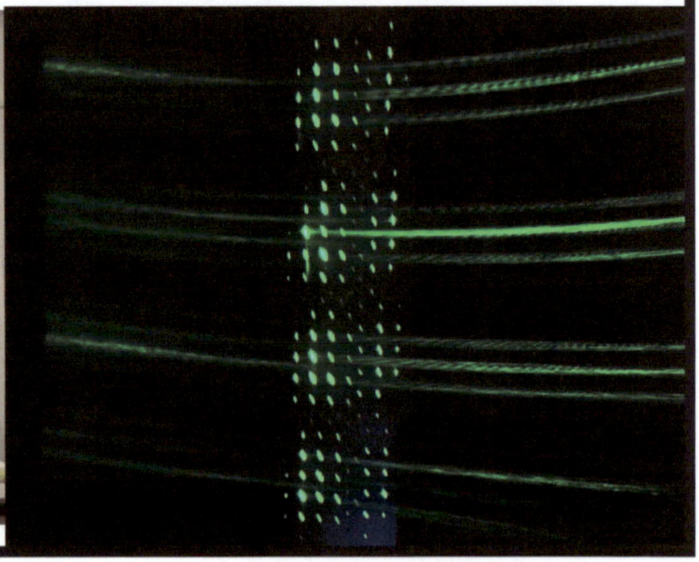

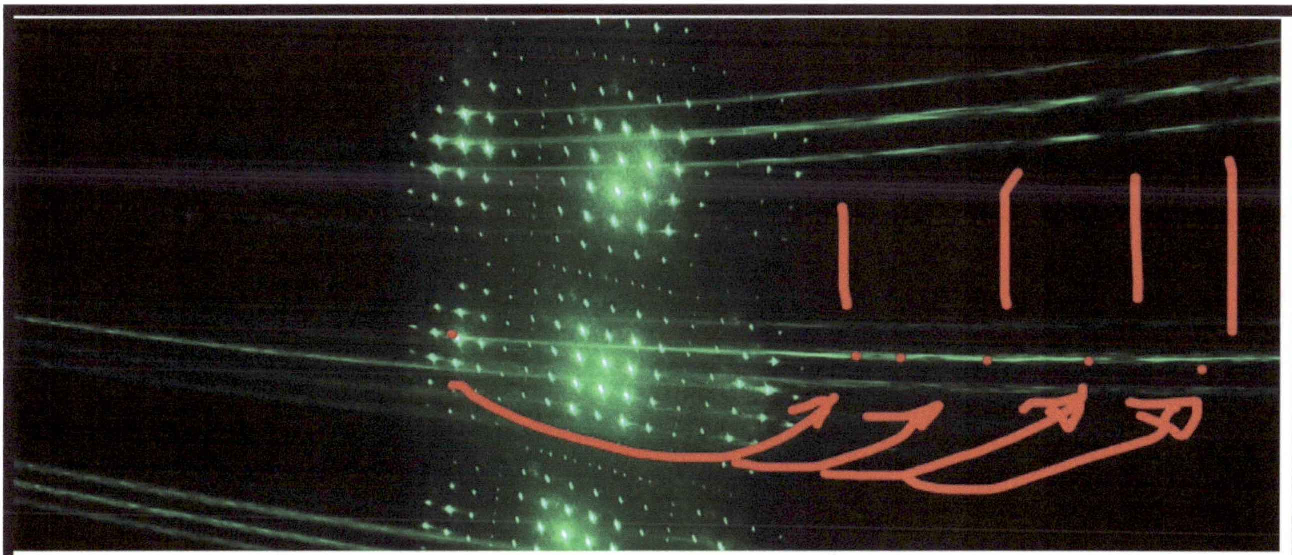

At a larger, zoomed scale of the light sources and cylindrical mirrors we obtain a scaled image identical to the one in the DSE, only reflections without any interference.

We can see on the wall how the image continues, but to see better we will change the distances and shooting angles to recover more light.

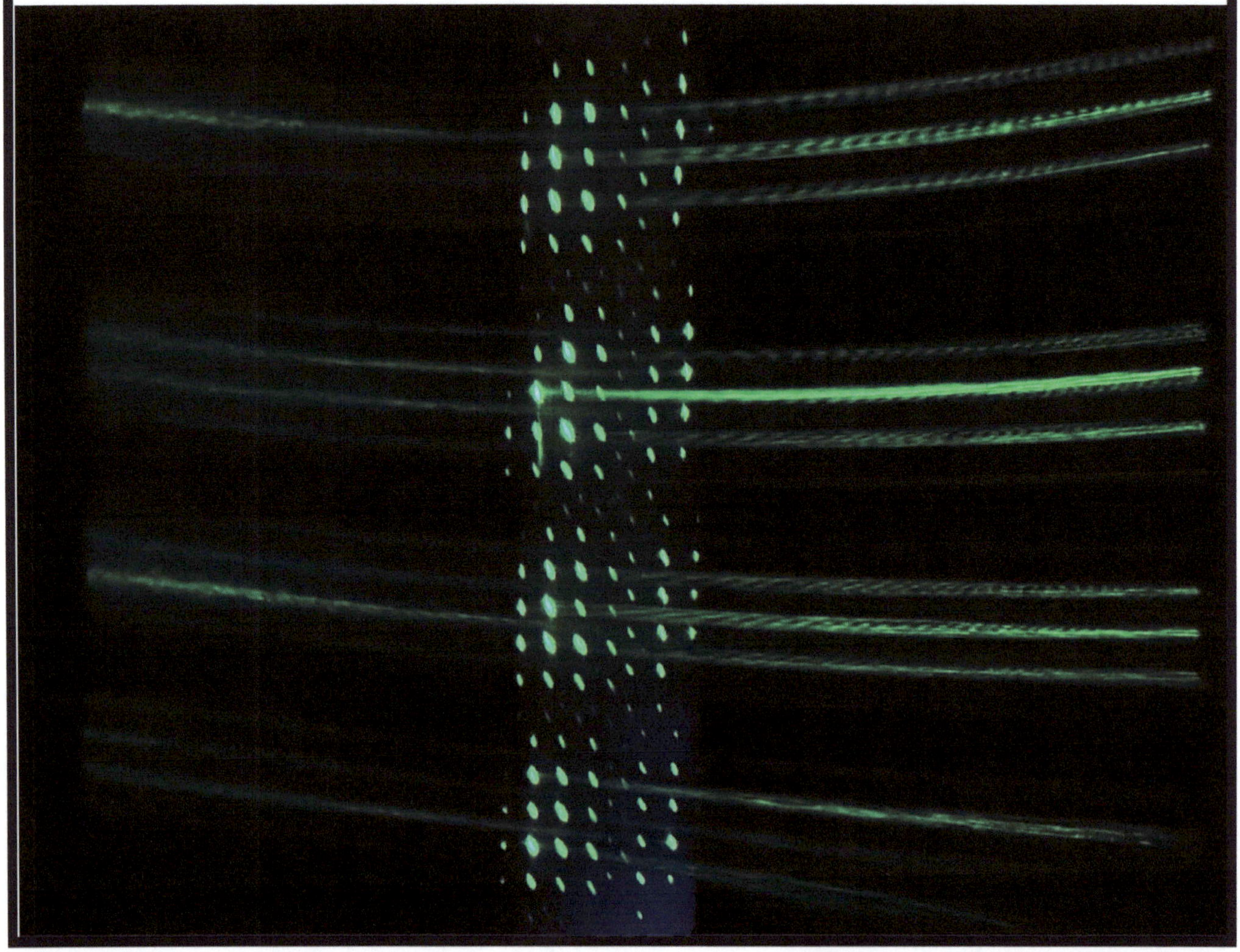

And tilting it obtains separate overlapping images of each "z"

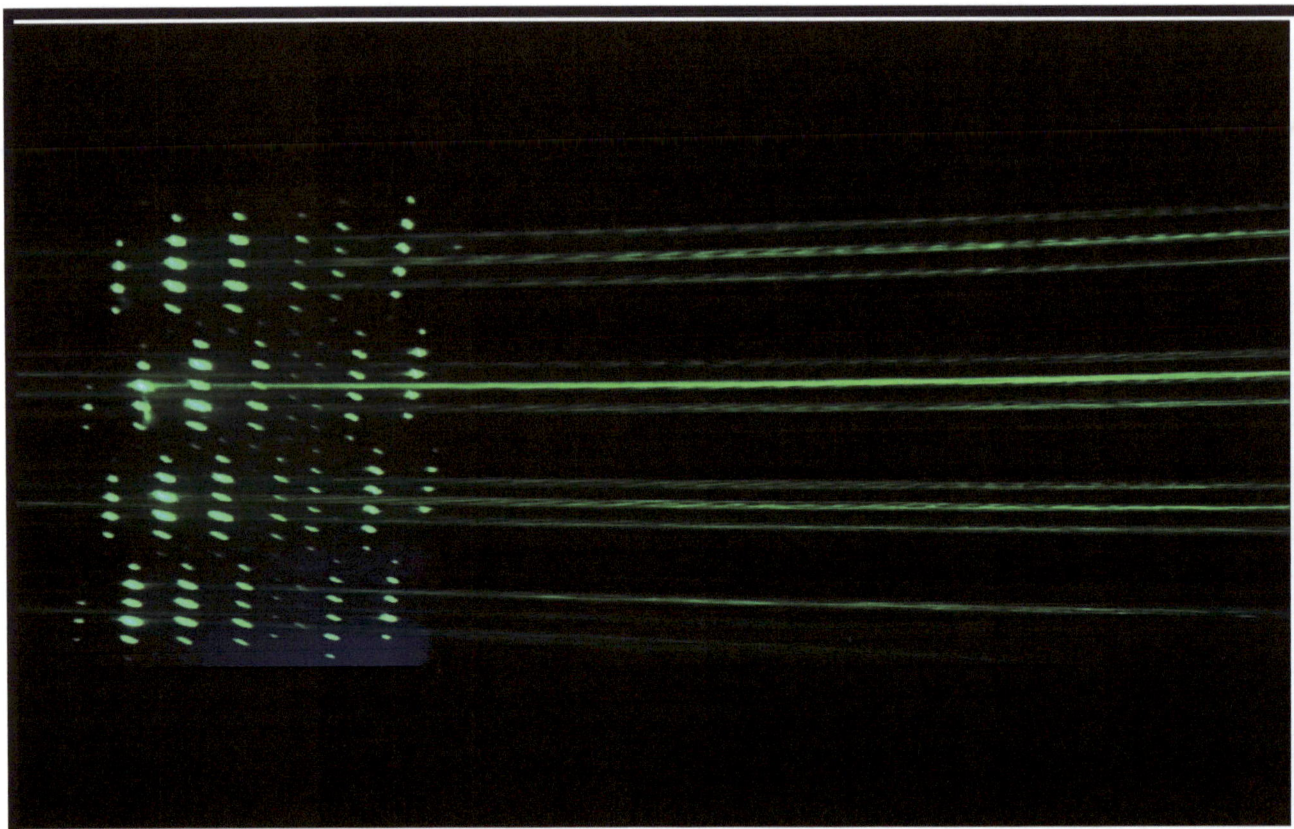

When we zoom in ,we notice that the bright spots on the wall are formed by the overlapping of different reflections of the 'z's (4 arms of z in each image) and that the distance between them corresponds to the distance between the light sources (2 cm)

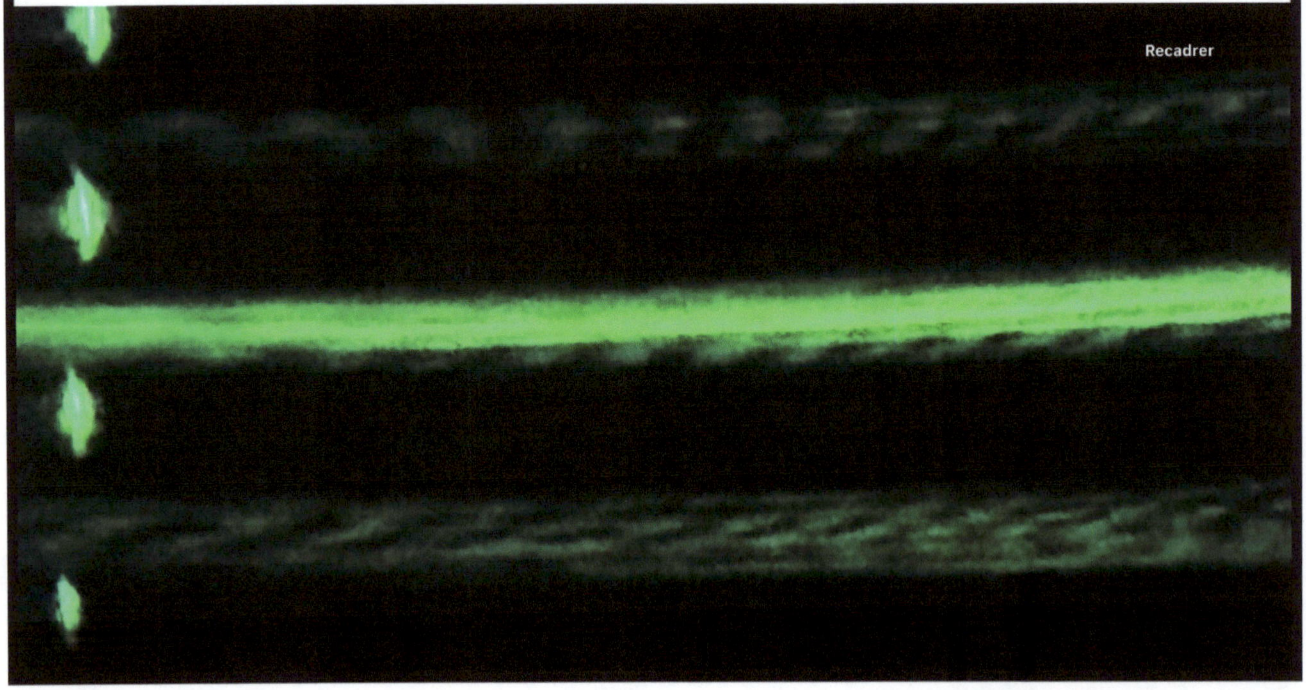

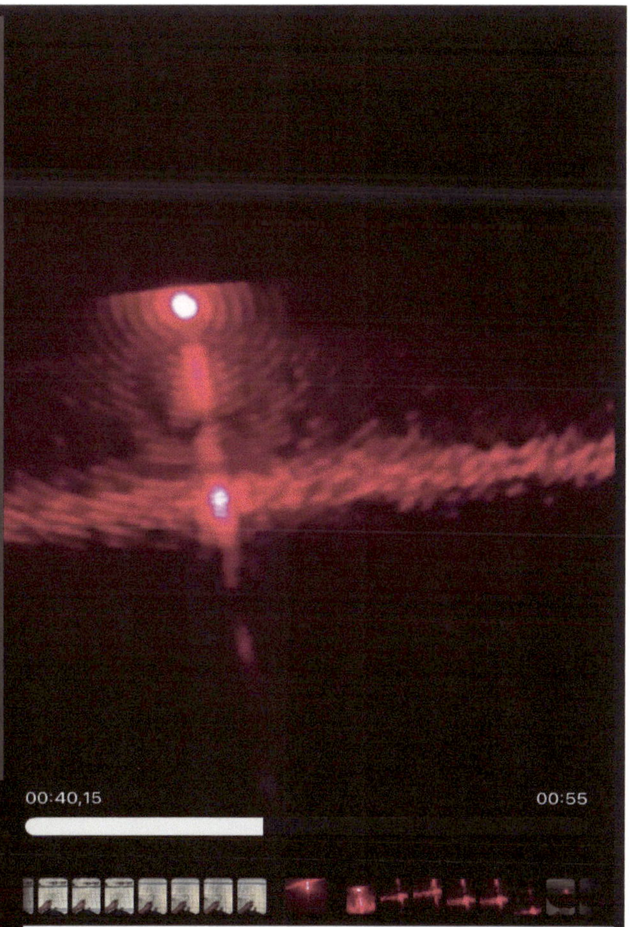

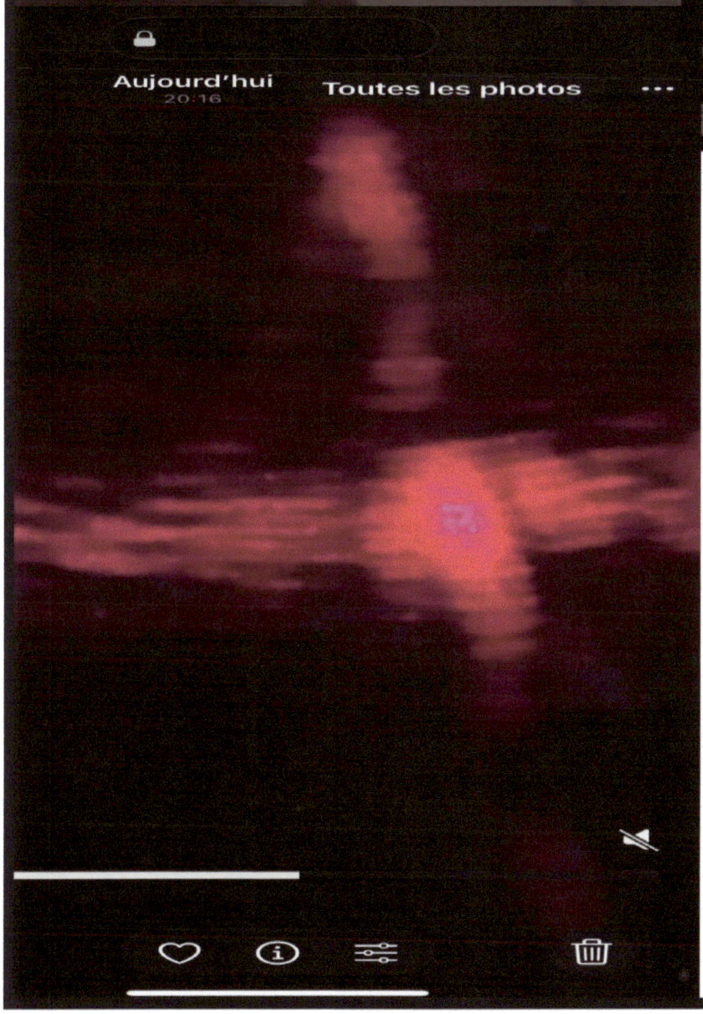

Using the grooved stainless steel from a spatula we can separate the z arms from a real DSE experiment. A kind of Electrophoresis of a single spot from DSE (I'm a doctor), and to answer the stupid question which hole the photon passed through, in fact WHICH MIRROR WAS THE LAST REFLECTED ON

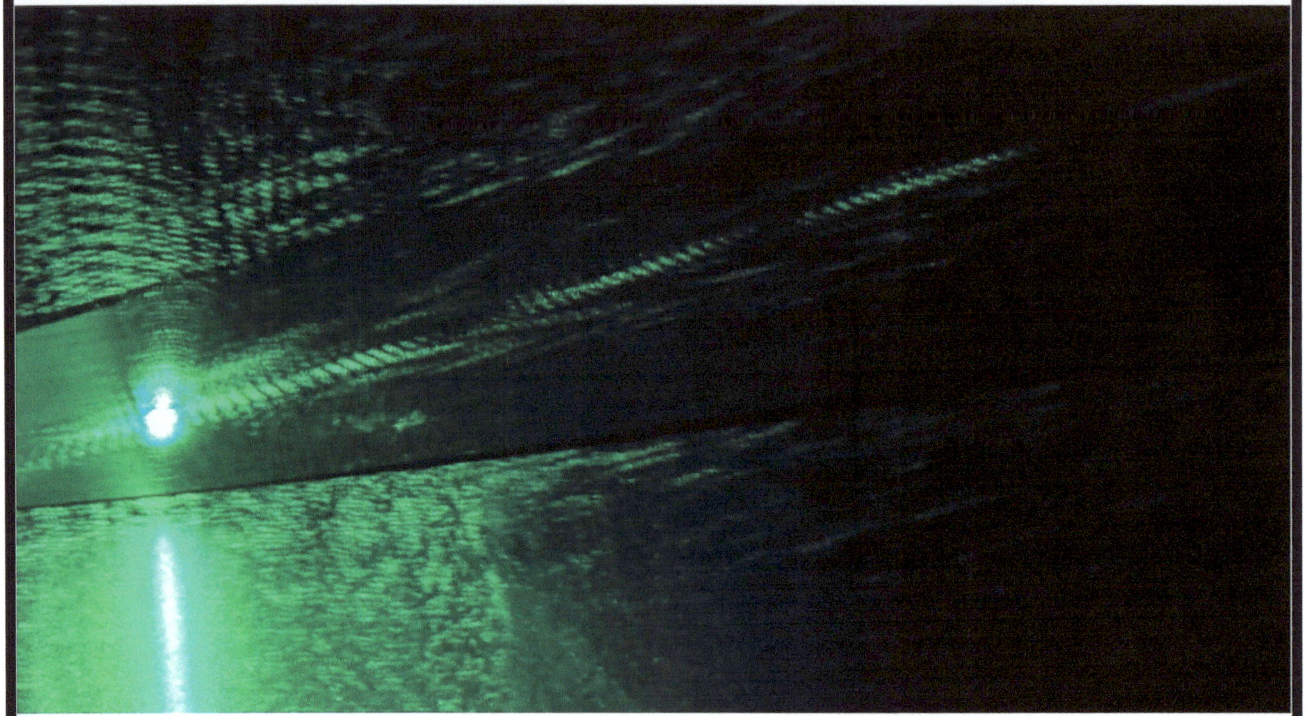

On a knife with finer striations we can do the same thing for green light, when we tilt and twist the wires, and separate each individual quanta that will form the images from the different reflection arms of the "z"s

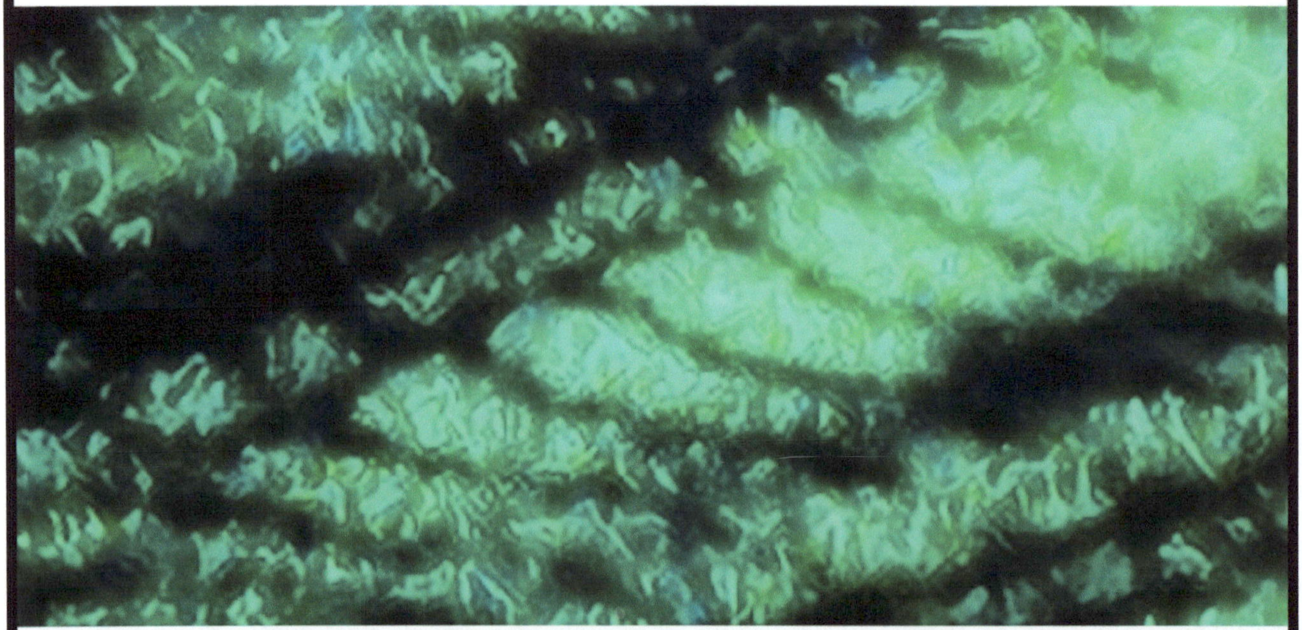

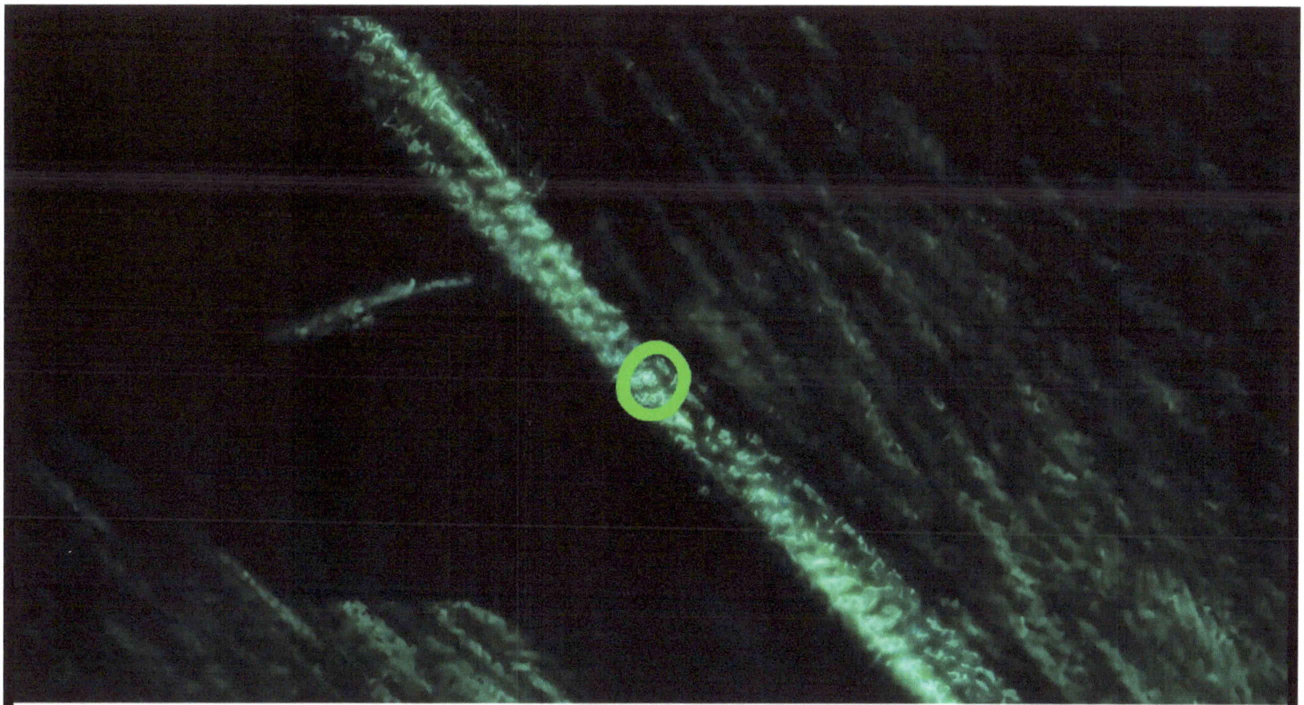

And we can Electrophoresis on a stainless steel knife miner at each point in the DSE image. All the other lines that are seen in the image are lights reflected at angles that will not create an image on the wall, but here we can recover them using the metal mirror. They are part of the image of those large circles obtained on the hair on my head, it is noticeable that some points come from the left and others from the right, they are offset from both sides on the "electrophoresis" image

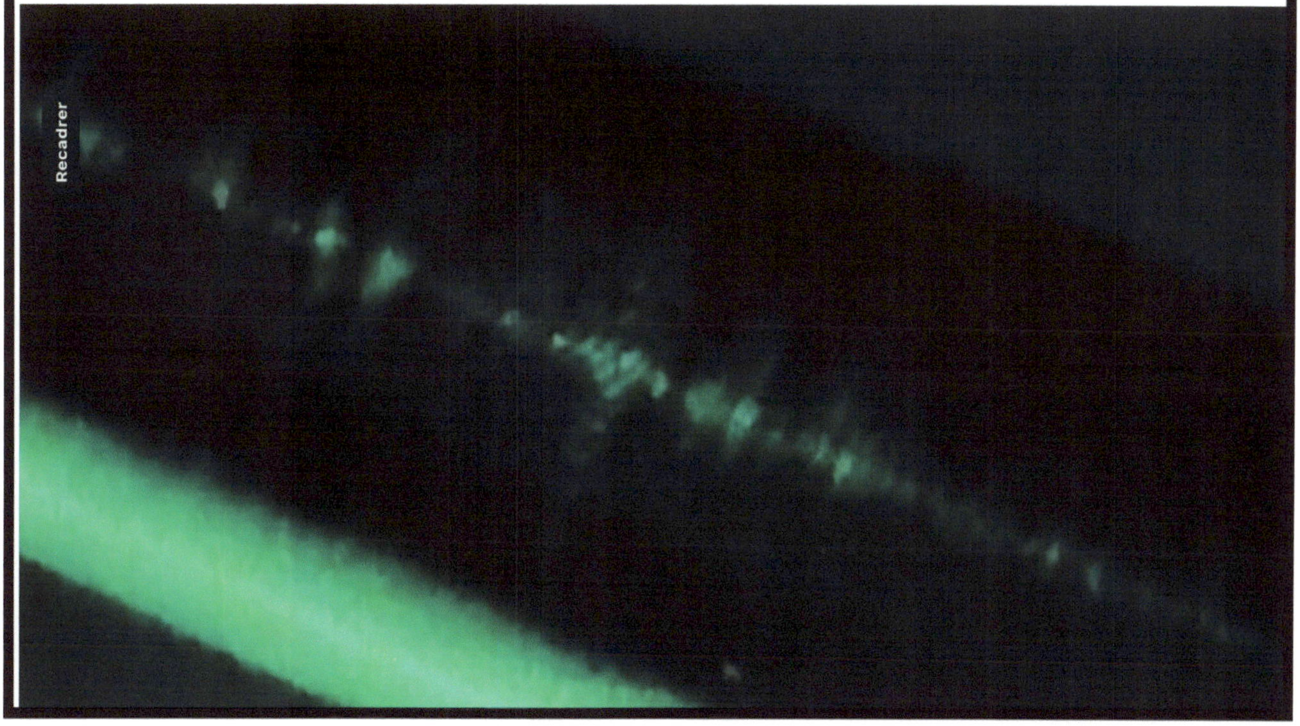

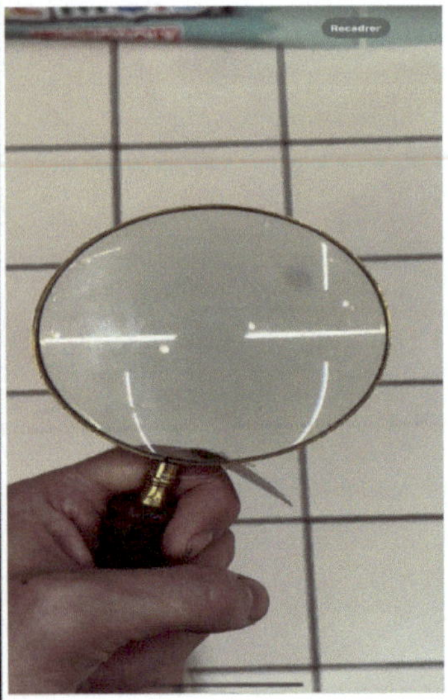

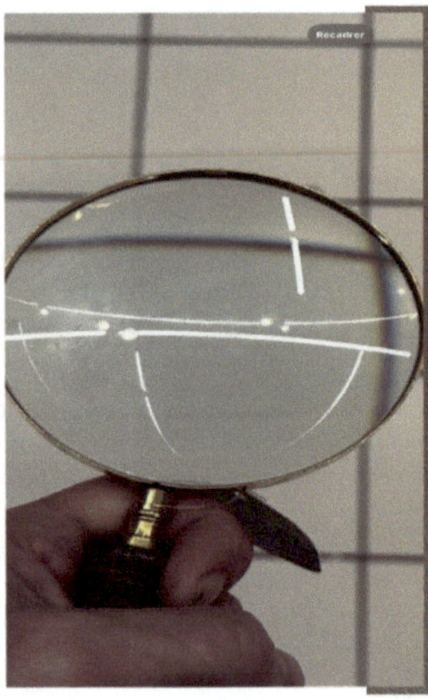

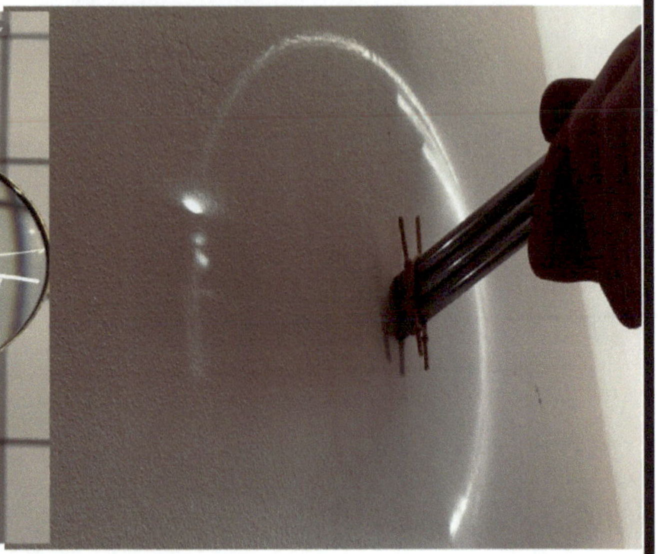

The" interference "patterns in images seen with the naked eye in nature, on a mosquito net, on monitor or television screens are not created by lenses, they are already in the image due to the light reflected on each group of atoms. Even if we change the shooting position and the lenses, this pattern does not change because it depends on the light source and the structure of the surface that reflects it.

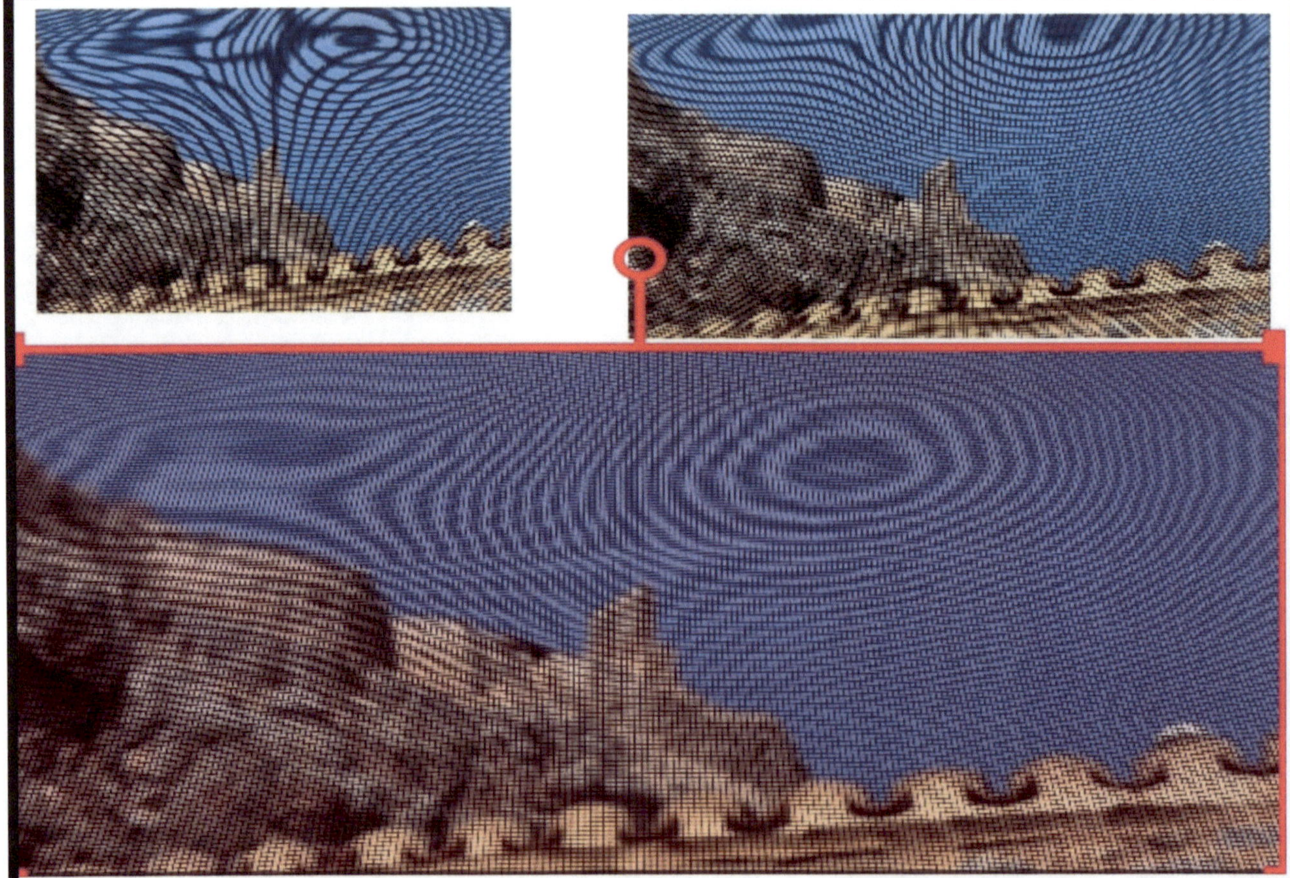

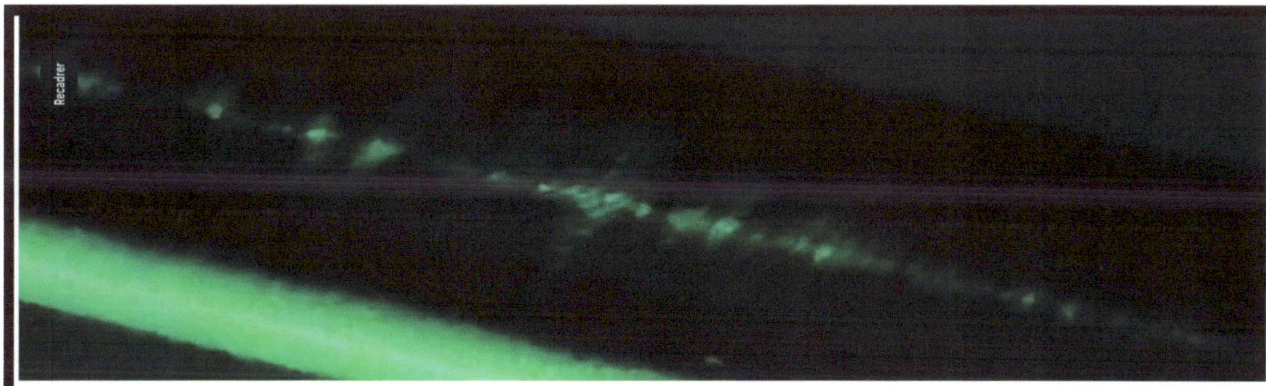

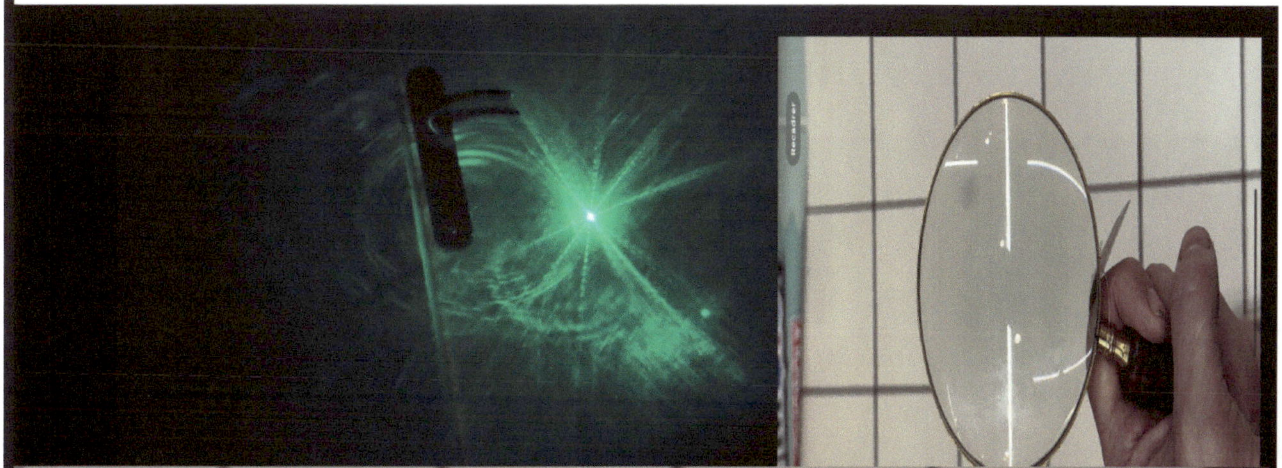

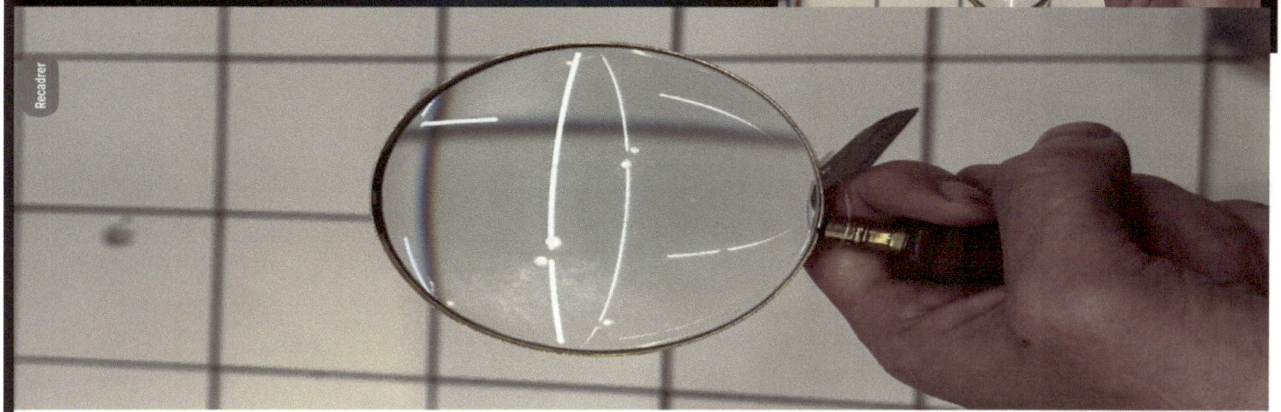

DSE as everything we see every second around us is just light reflected by groups of atoms

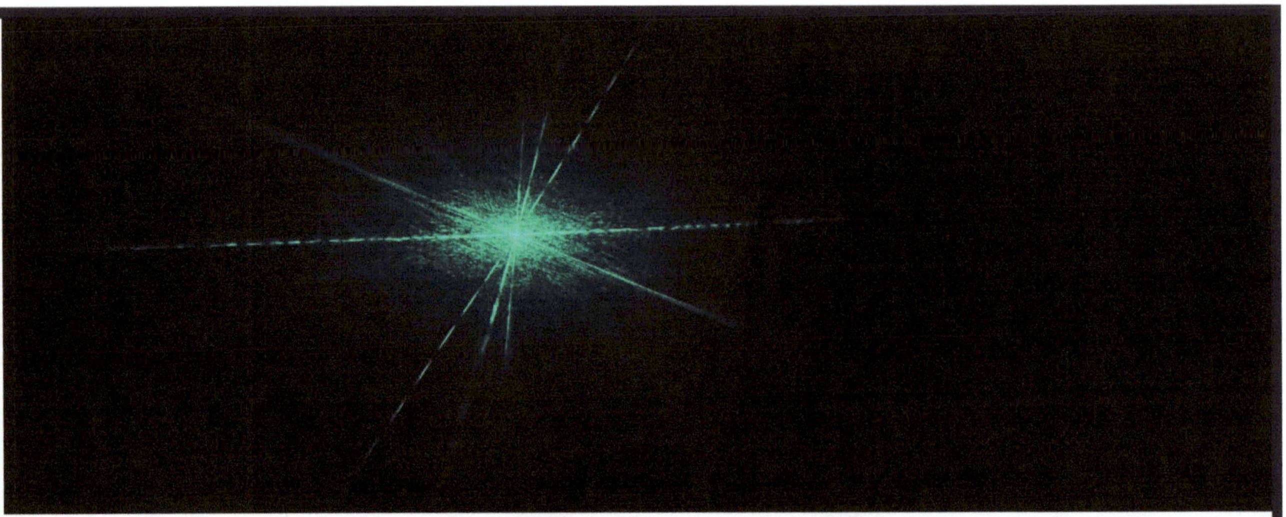

DSE with different sized slits

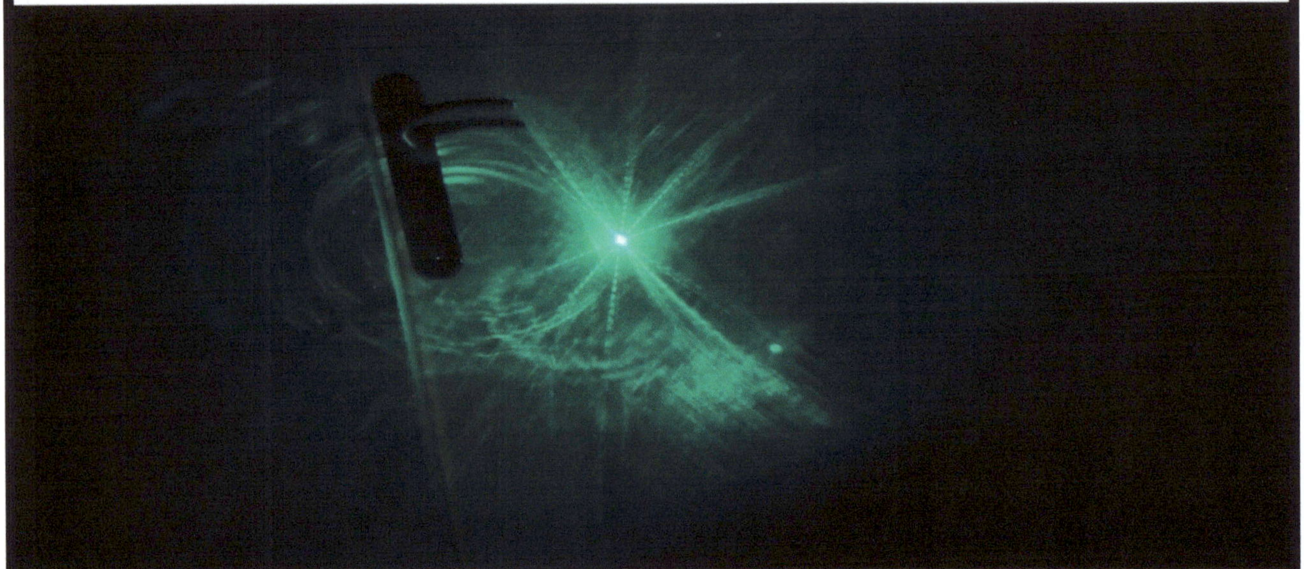

All these interference lines are located in a circle on the plane that passes through the slit in the DSE, first image it is an ideal situation in which we perfectly overlap all the "z"s from each point in the light source on the same line. Just optics and geometry.

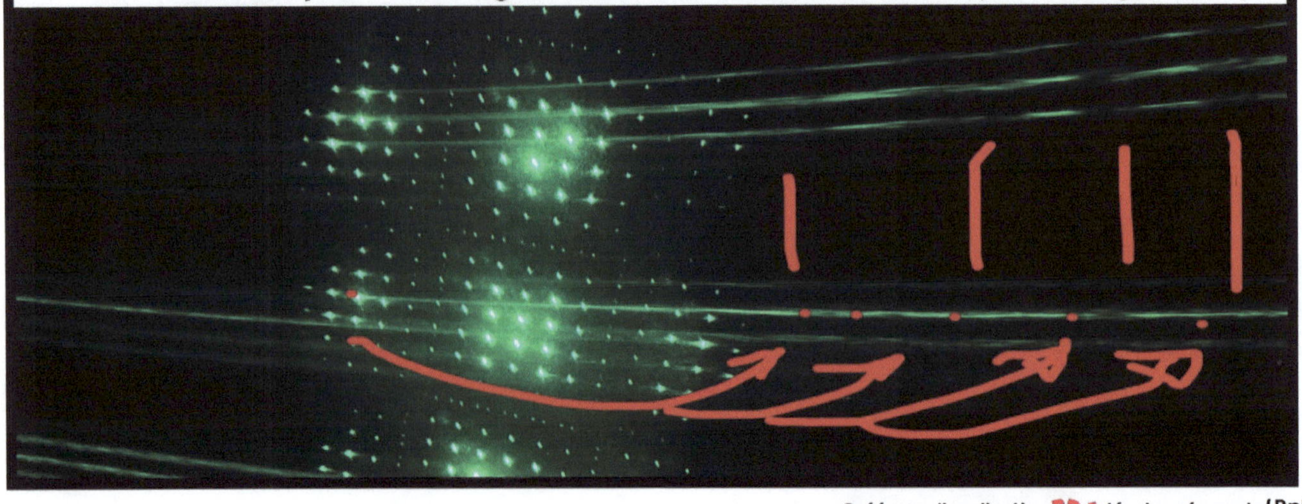

Without comments,3 cm wavelenght,DSE

It's beautiful,
isn't it,
identical to
DSE on 3 forks,
I ate DSE

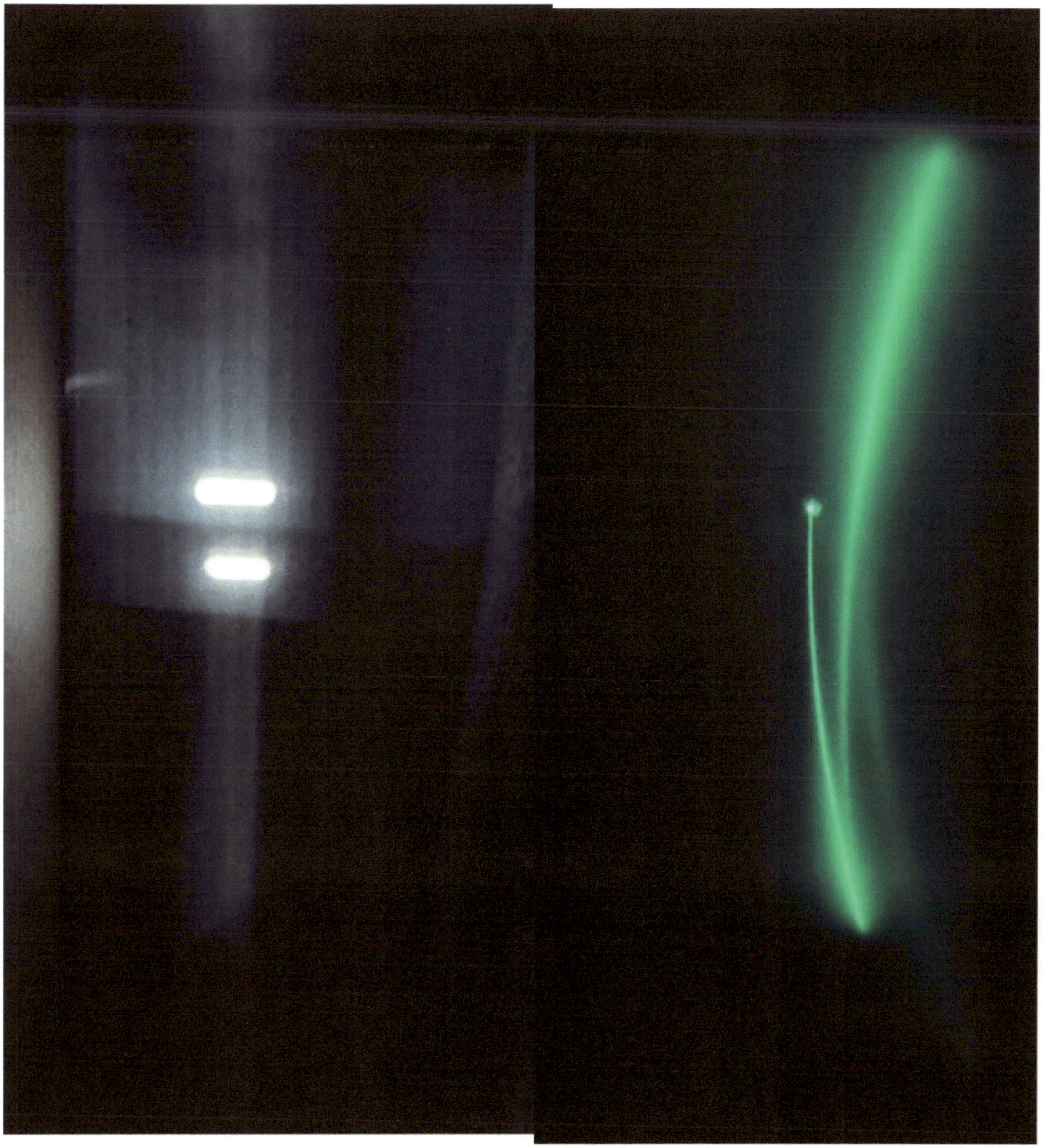

In full sun, nanometric wavelength we only see the main "z" superposed

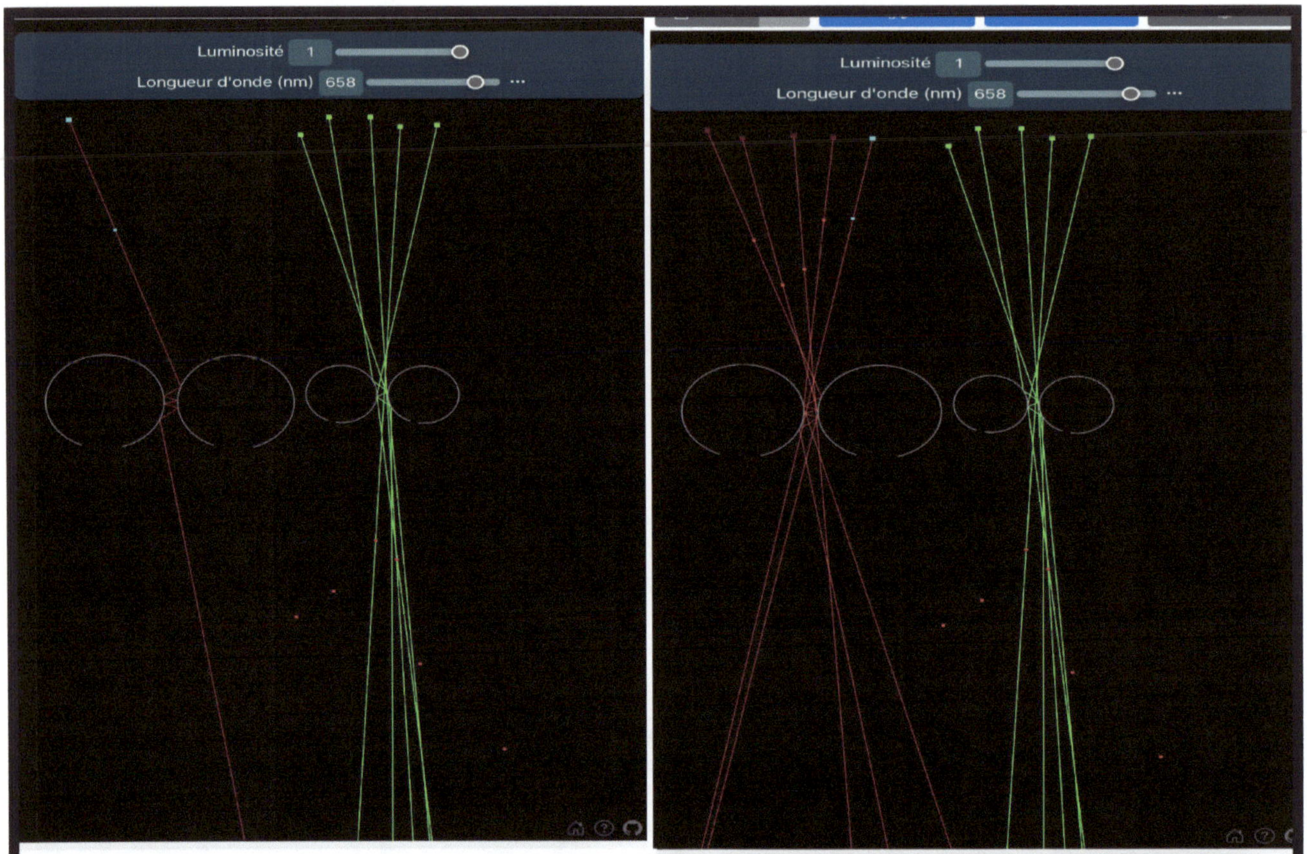

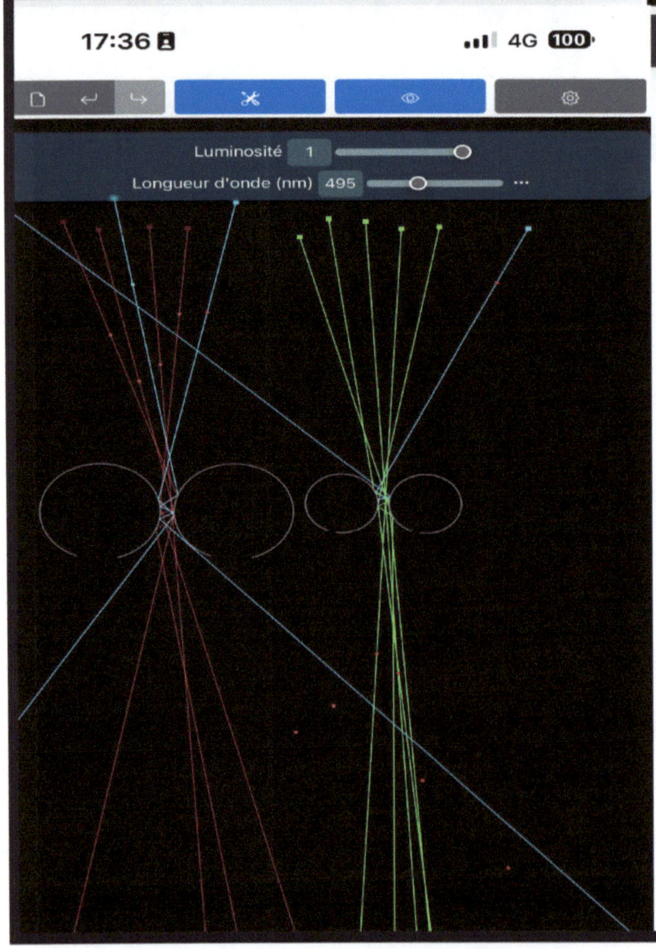

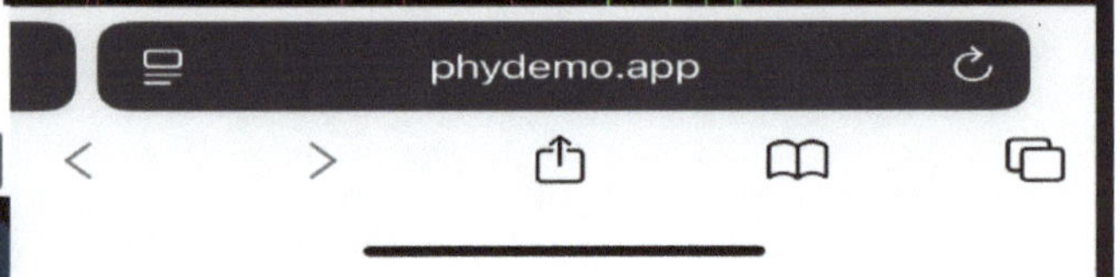

But for different wavelengths the reflection occurs at different distances from the spherical mirrors, the mirror will have a different radius for red or green light

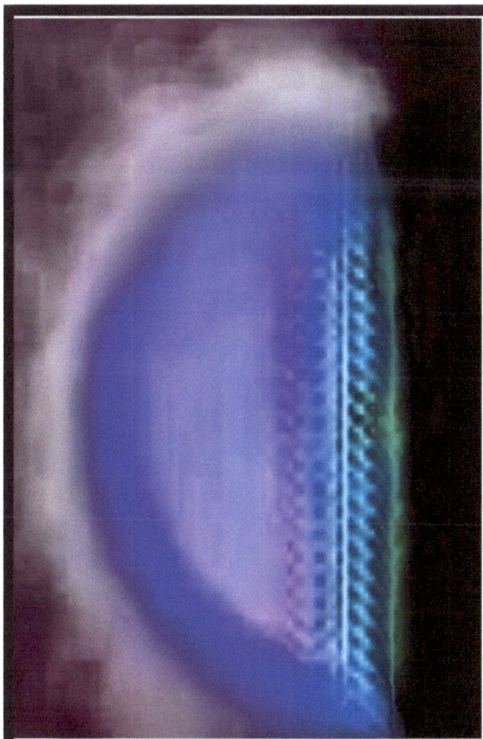

And we can observe this in the photograph of a lightning bolt from Wimhurst's Machine, or in various experiments with reflections on metallic mirrors.

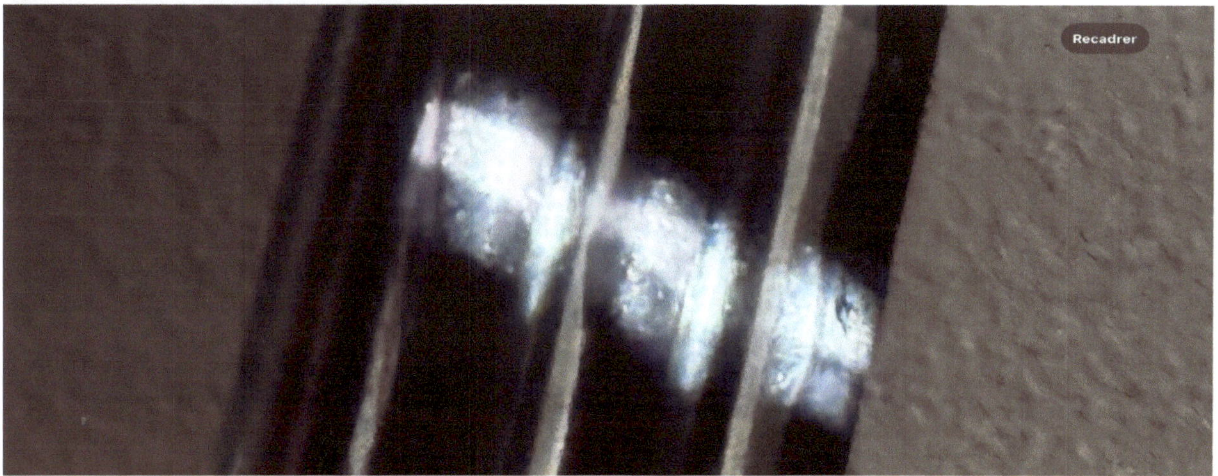

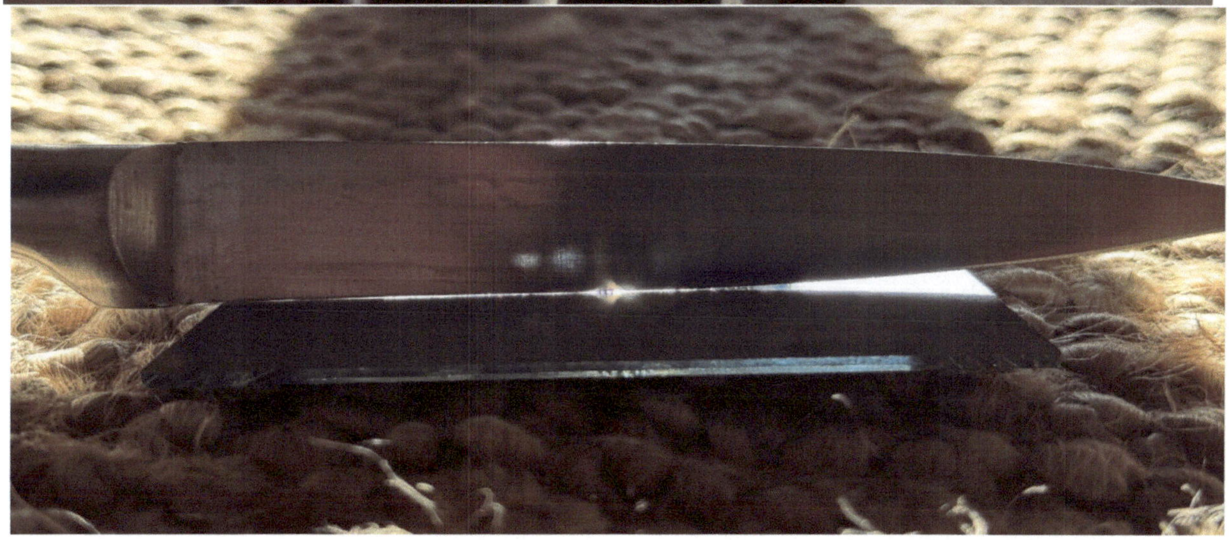

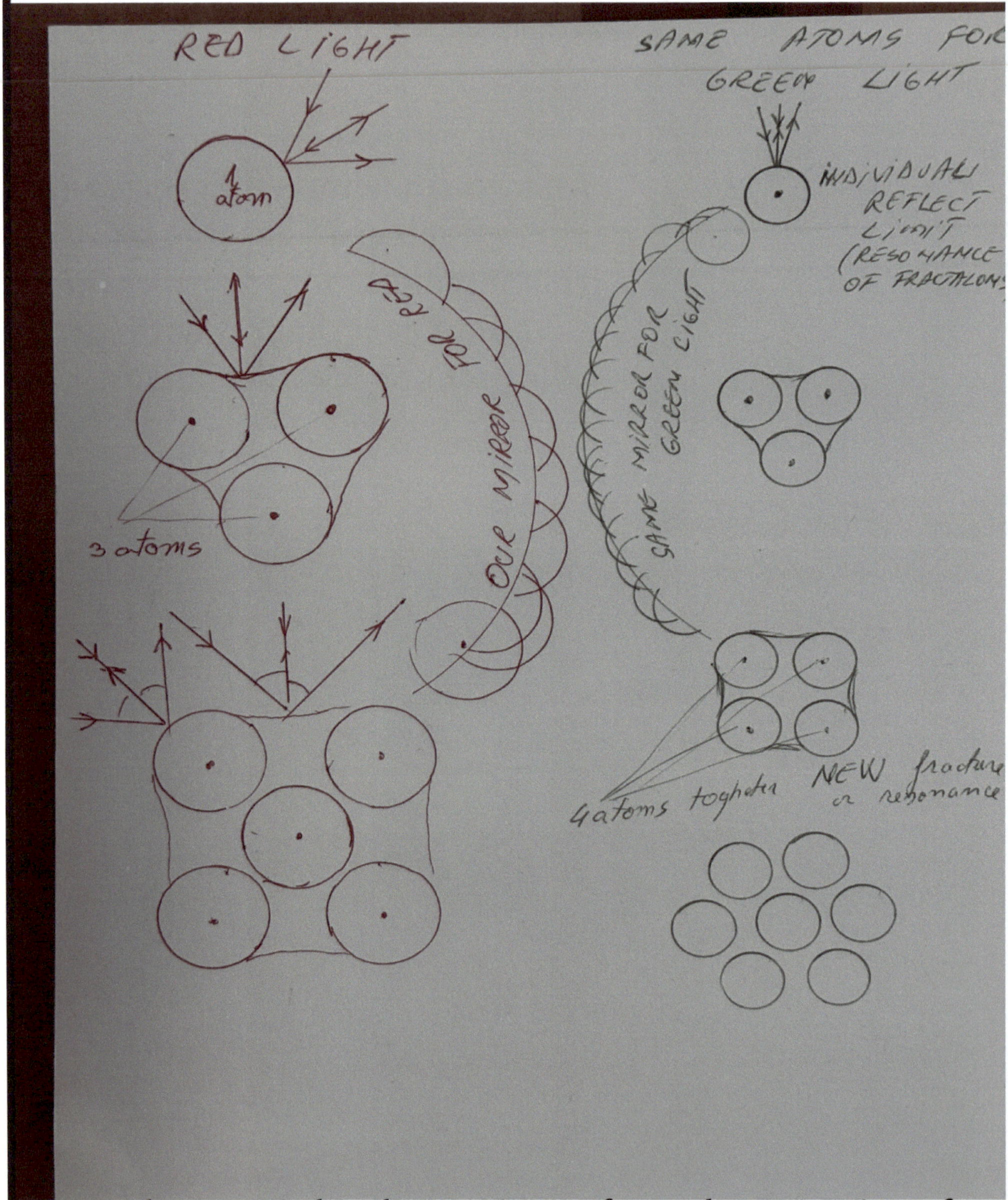

So at the atomic level our mirrors from the DSE wires, for red and green, look something like this, the atom is the tiny dot in the center of the circles

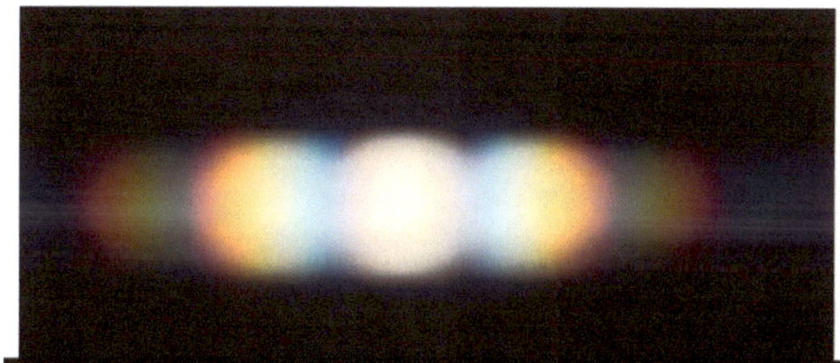

For that always RAINBOW wavelengths(and EM waves in general) are separated at different angles, it reflects at different distances from groups of atoms that form virtual mirrors in the ether

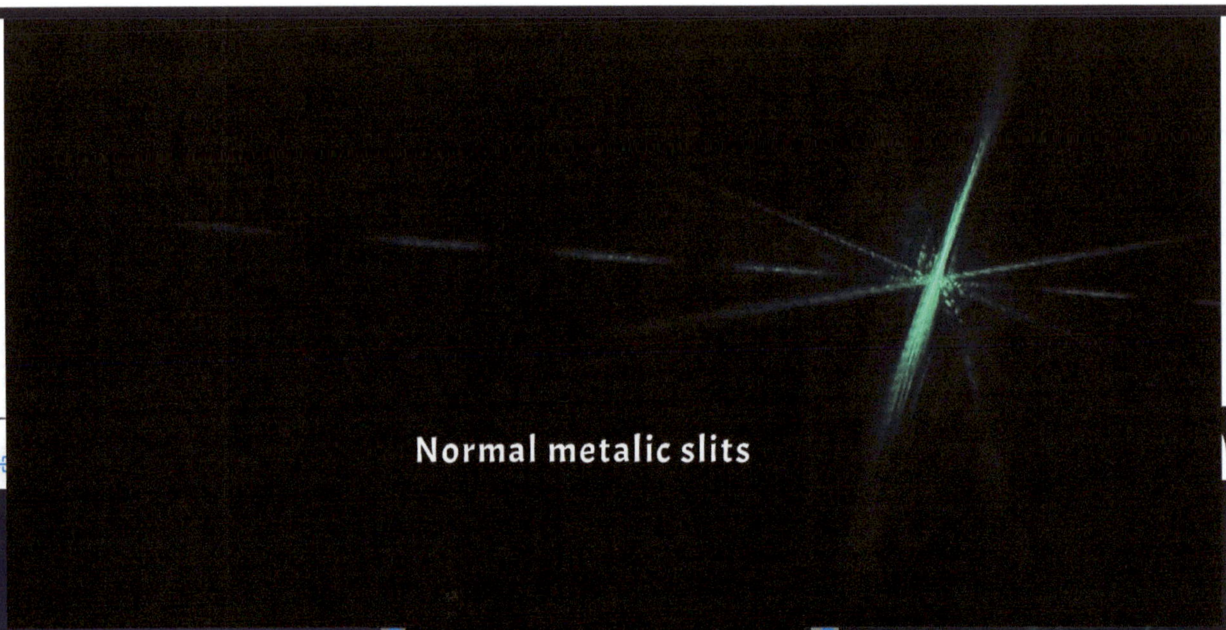

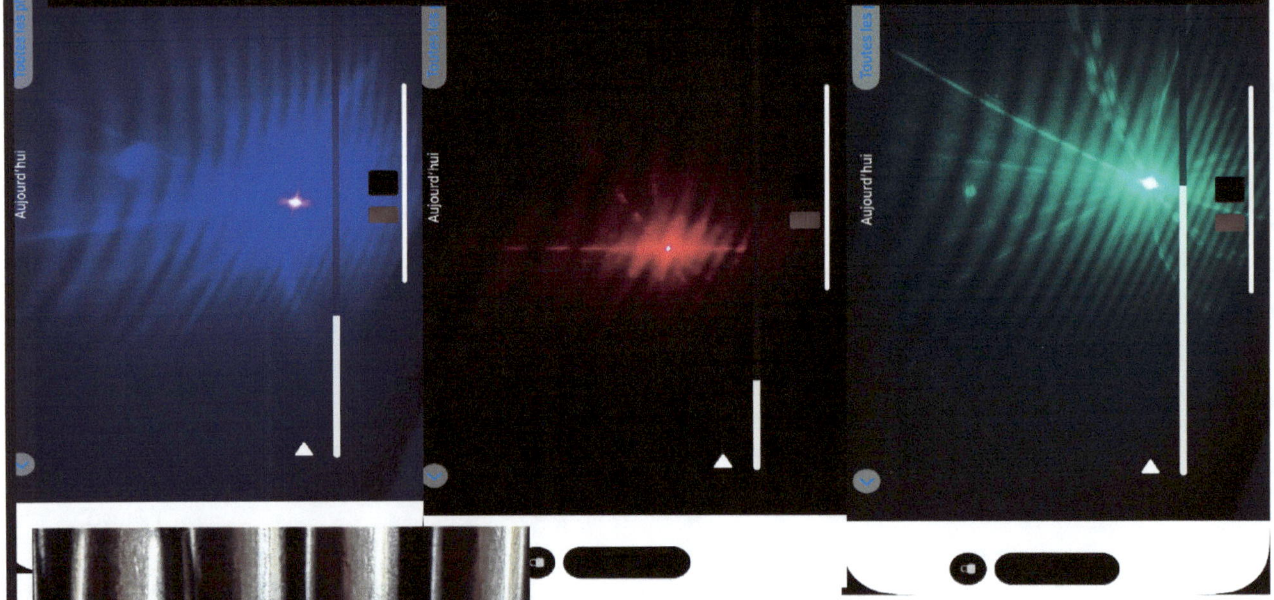

On a special plastic mirror we can see the specific pattern of each wavelength that will correspond to the distance between the bright points, and will be found in all the grooves, the slits, because it depends on the structure and organization of the atoms. Photos with the camera ,mirror and laser position identically, we see that the number of slits varies with the wavelength

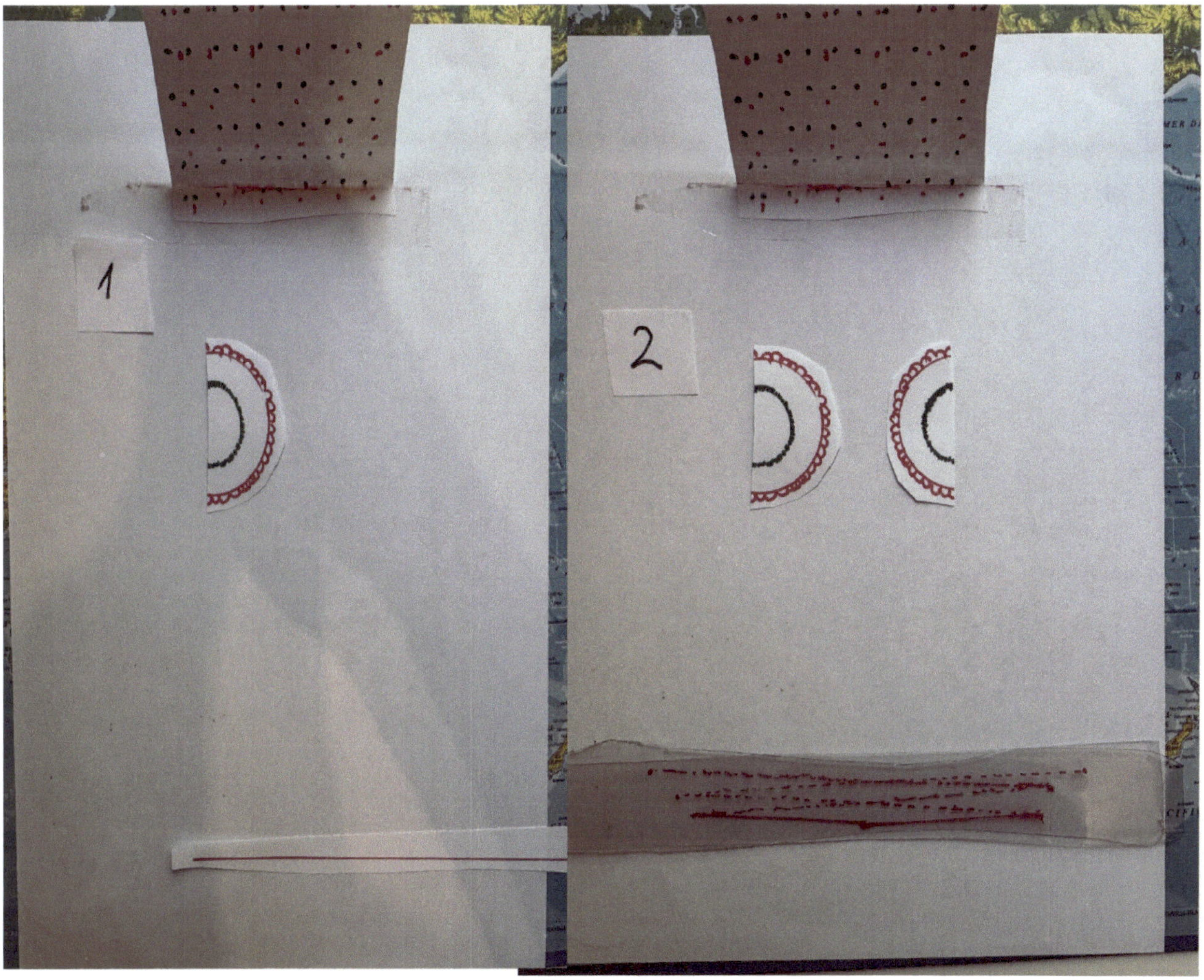

Decomposition of image formation into stages in DSE, one half of a mirror and his image , then two halves with the appearance of "Z"s

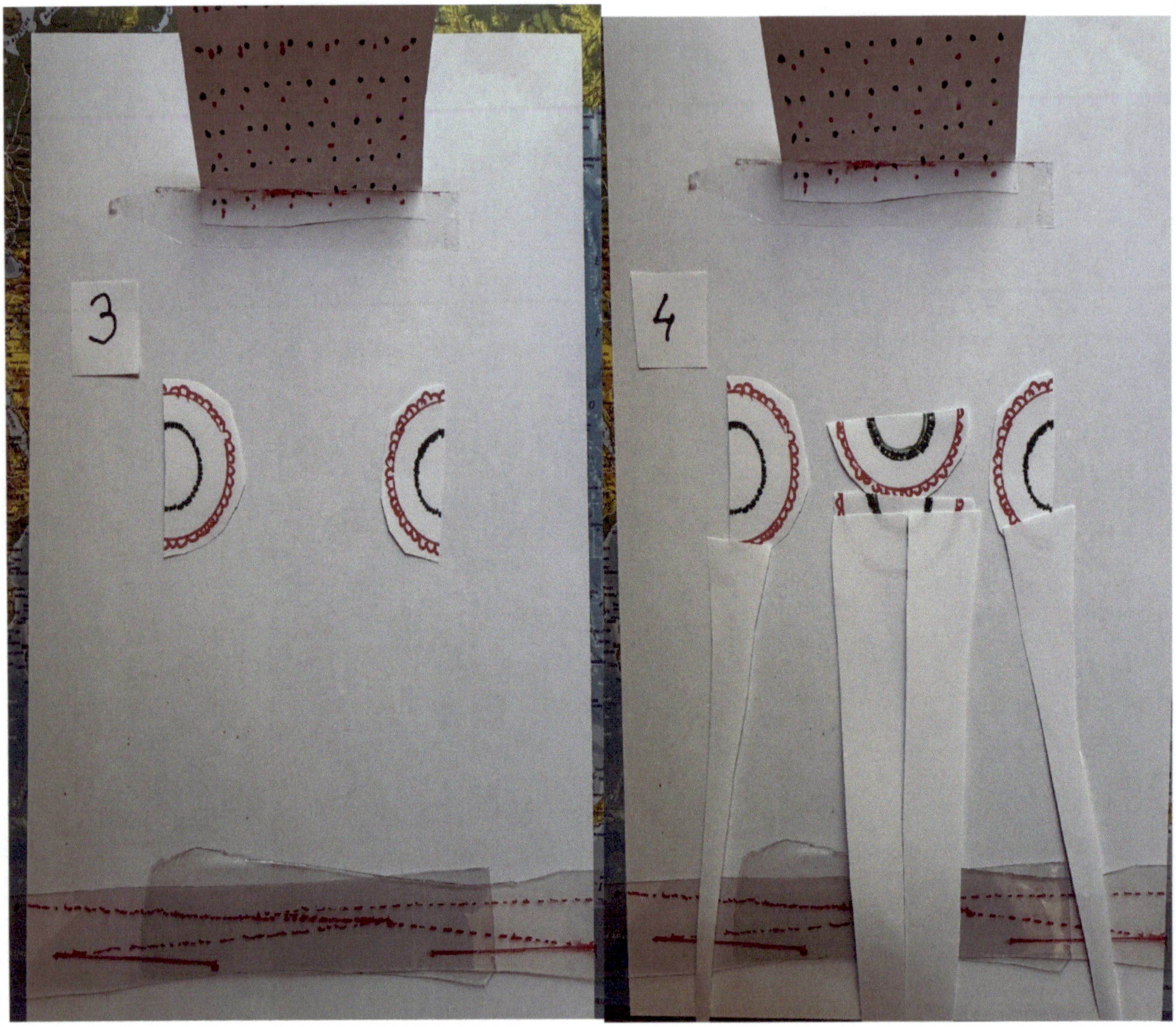

Separating the two halves with the simultaneous movement of the "z"s on the wall by an equivalent distance, and then the shadows from the added mirror and from the first two mirror halves

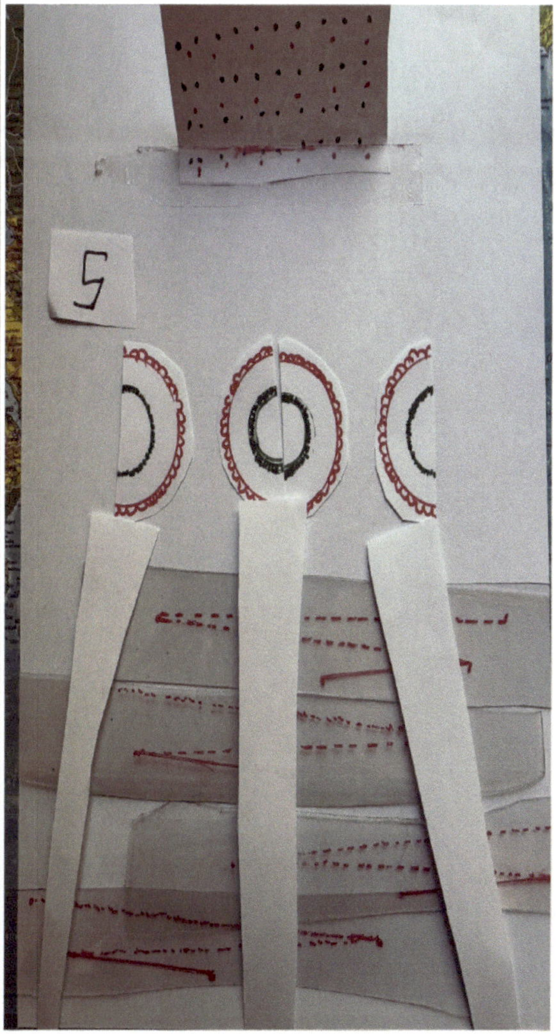

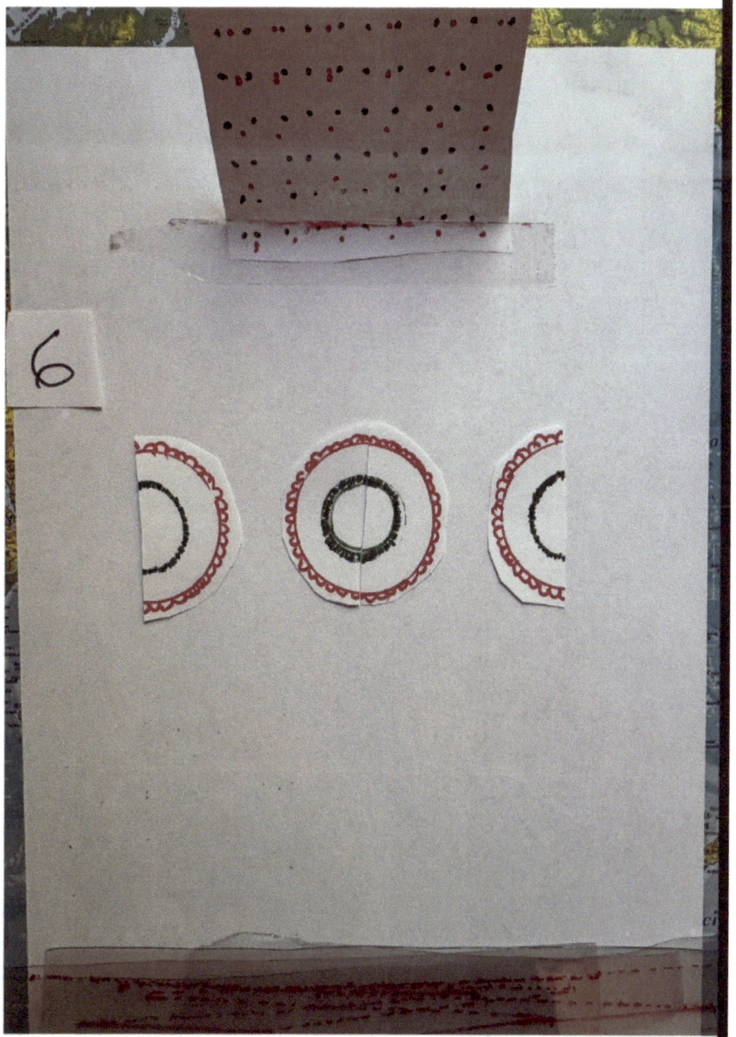

And finally by superposing the unique reflected points on each possible z we obtain the final image, we must understand that at the atomic level each wavelength is reflected on the principle of all or nothing, it is never divided into two parts, the virtual mirrors that are formed in the ether are a kind of concave mirrors with a diameter perfectly equal to the wavelength and are created by the light wave, by Schrodinger's balloon of fractalons.

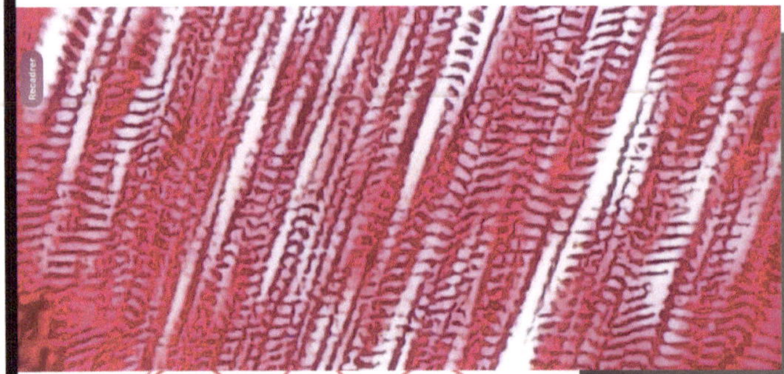

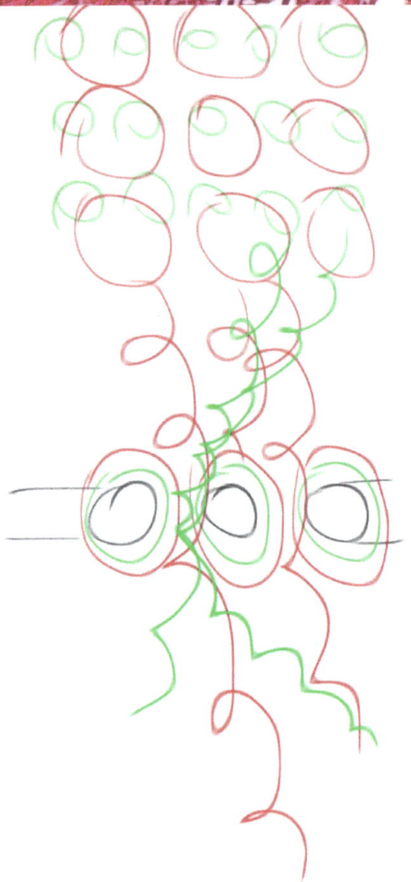

We will have a different position of the "z" for green and red, if we simulate shadows with 20 light sources (holes in a paper and light from the sun, and "glued forks") we understand how the image is formed for each wavelength

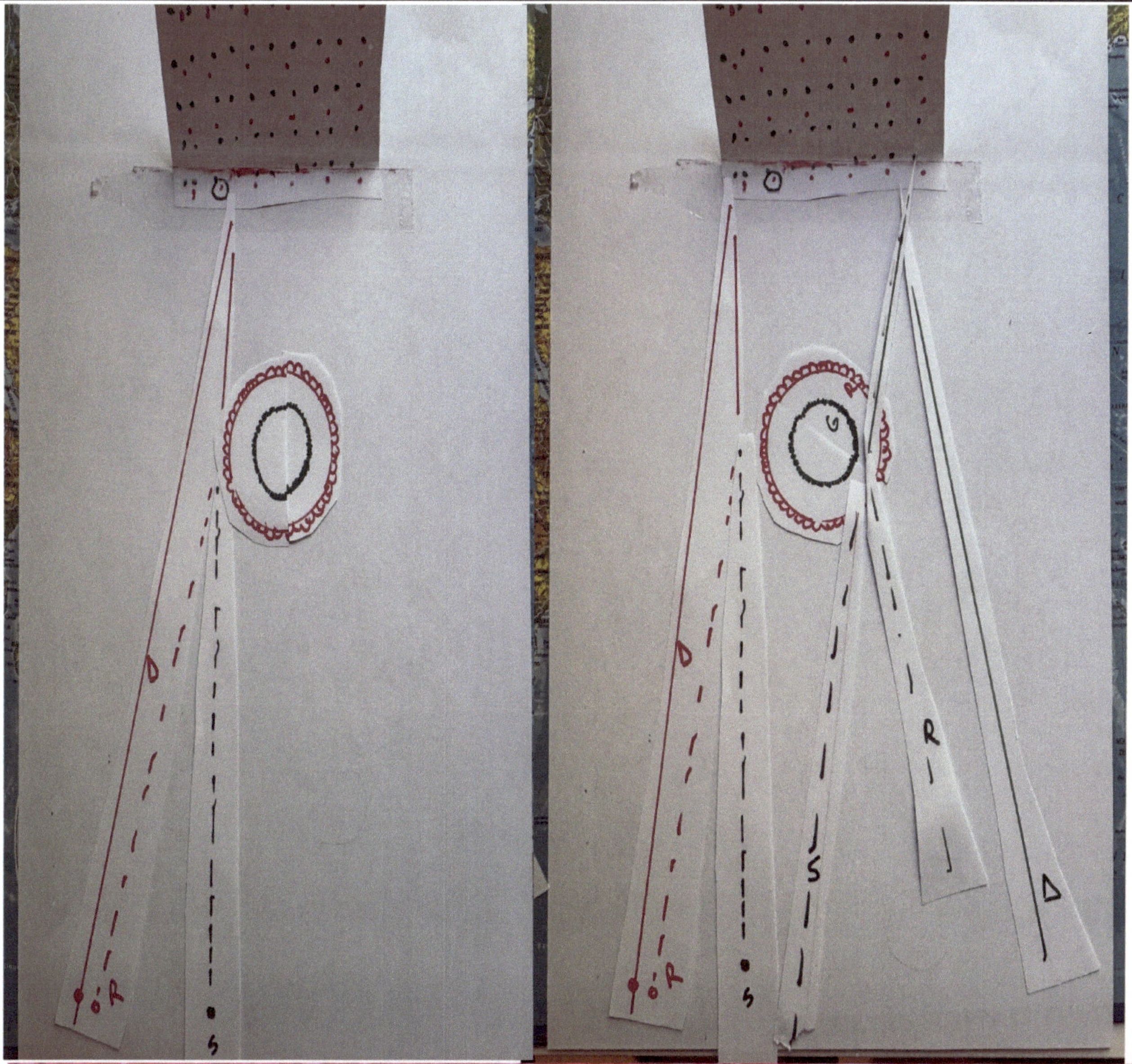

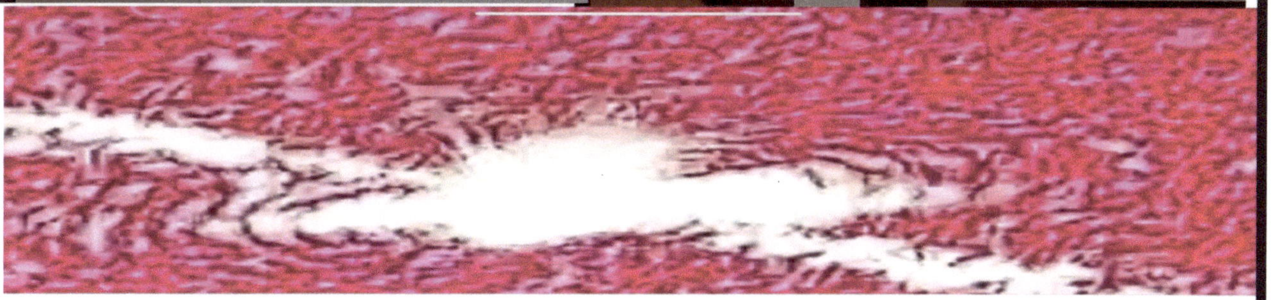

When the observer looks directly at the slits in the DSE, from a lighted point on the wall, he will see a virtual image in the continuation of the reflected ray, he will have the impression that the light source is in other positions, different for each reflected ray, and for each color.

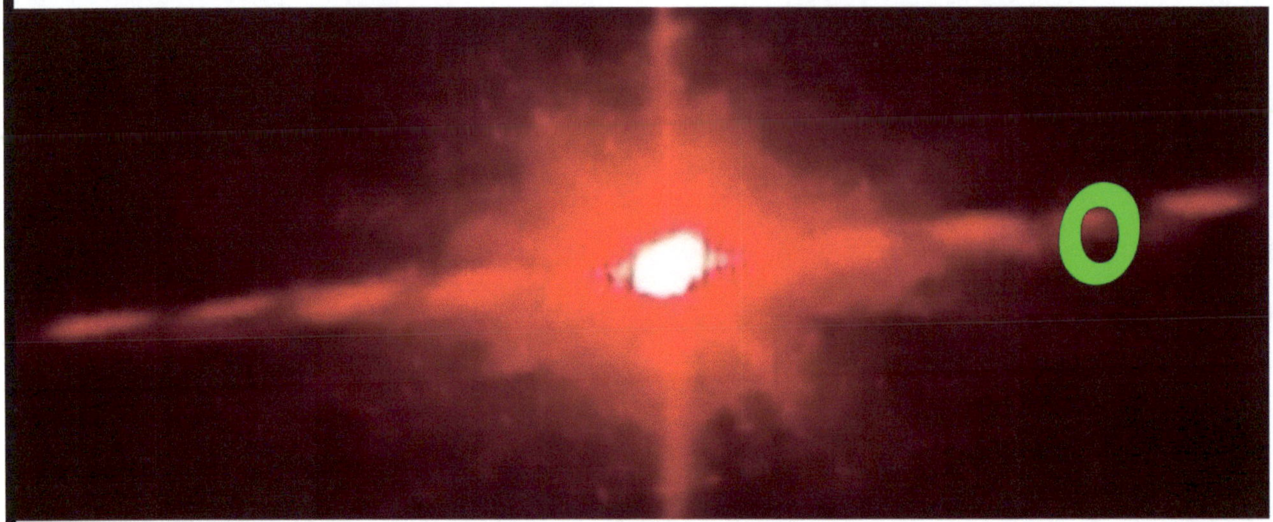

Classic DSE with the image projected on the wall 2 m away from the slots

These two pictures demonstrate that the image in the DSE is a mirror image, if it were an interference image in the second picture we should have a single bright point

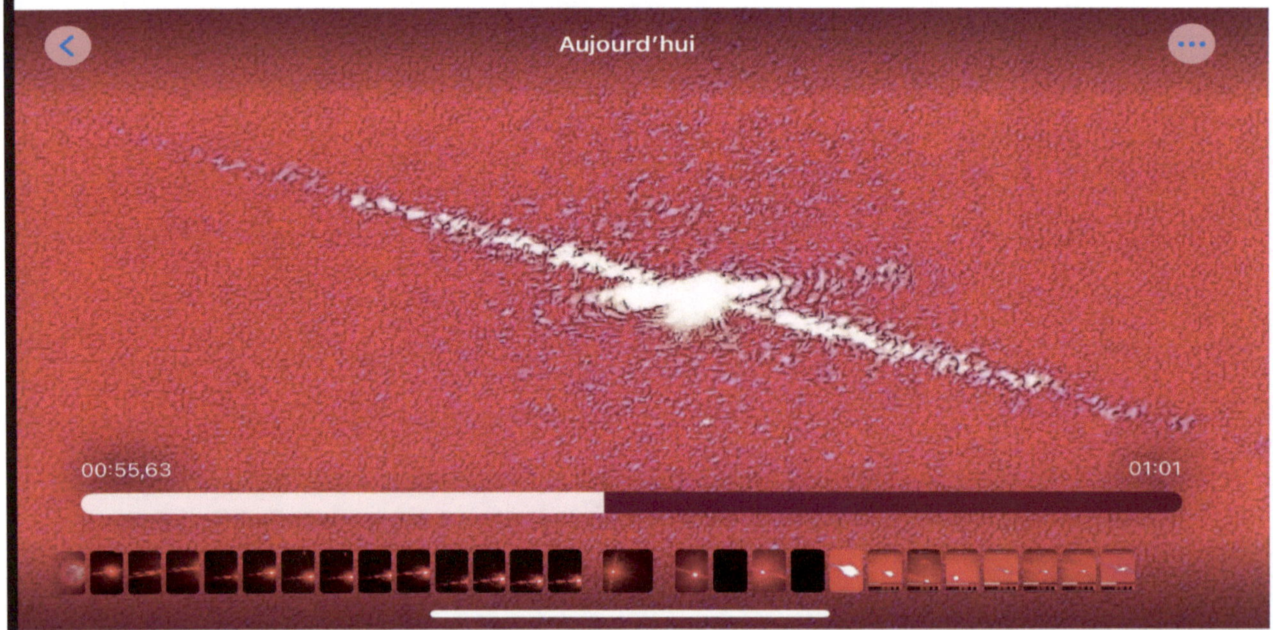

Picture taken of the two wires with the iPhone zoom x5 from the green dot

We must imagine each energy quanta similar to a spring, all 5 Schrodinger balloons (wavelenght quanta , in actual physics photons)will manage to pass through the large hole by reflection(in the bottom images) , reflection occurs at the interface with the vacuum not with matter, I will explain later

Depending on the width of the slit, some wavelengths can pass, others cannot, and each will be reflected at a different angle - RAINBOW, because of this ,the things around have different colors, the fractal structure on the periphery of the atoms will allow the "DSE like" reflection of only certain wavelengths.

In these images we can see that different wavelengths are reflected on the cylindrical mirrors(represented by the mosquito net)at the same angle, practically the rainbow that is seen is a projection of the sun in the lateral cylindrical mirrors, we see part of the sun shifted on different wavelengths in the mirrors formed by the lines and columns of the mosquito net(diameter for each wavelendght mirror is D+2x wavelenght)

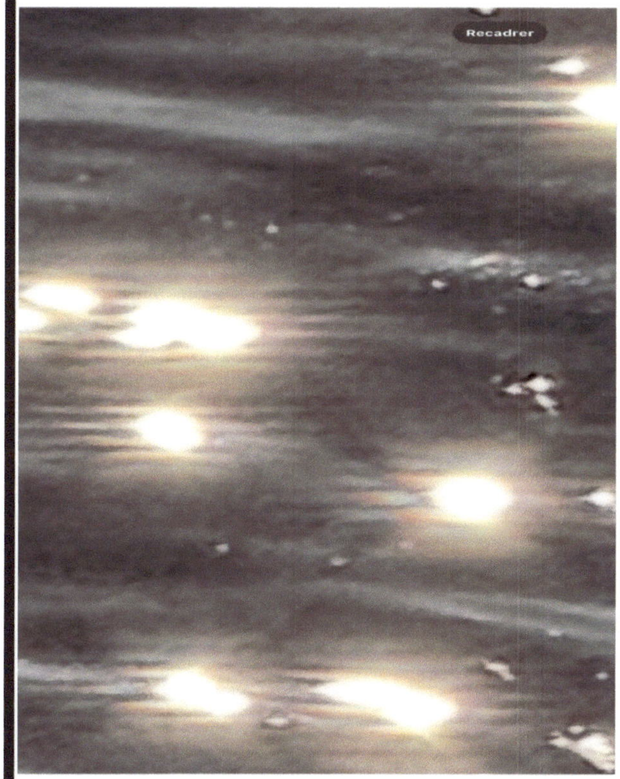

The same phenomenon on sea waves, we see the sun in several cylindrical mirrors represented by waves, each wavelength being reflected at a different angle, 2x the wavelength is added to the diameter of the wave

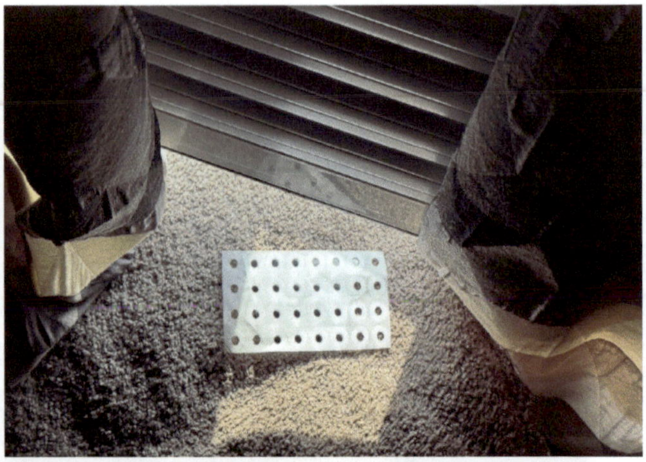

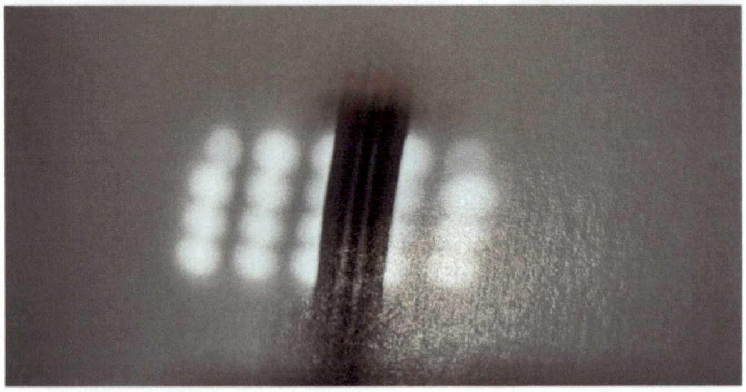

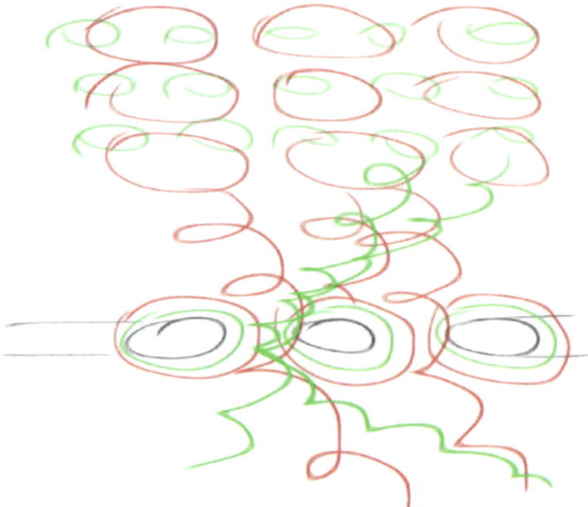

To better understand, we need to imagine the light source made up of several smaller light sources, exactly like the two pictures taken at the DSE wires in red and green. In the bottom picture we have a reflection image on 3 cylindrical mirrors with a diameter of 2 cm, in the top picture with a diameter of 0.5 cm, just shadow, no "interference" .The light source is a mirror in the sun with a perforated paper on it.

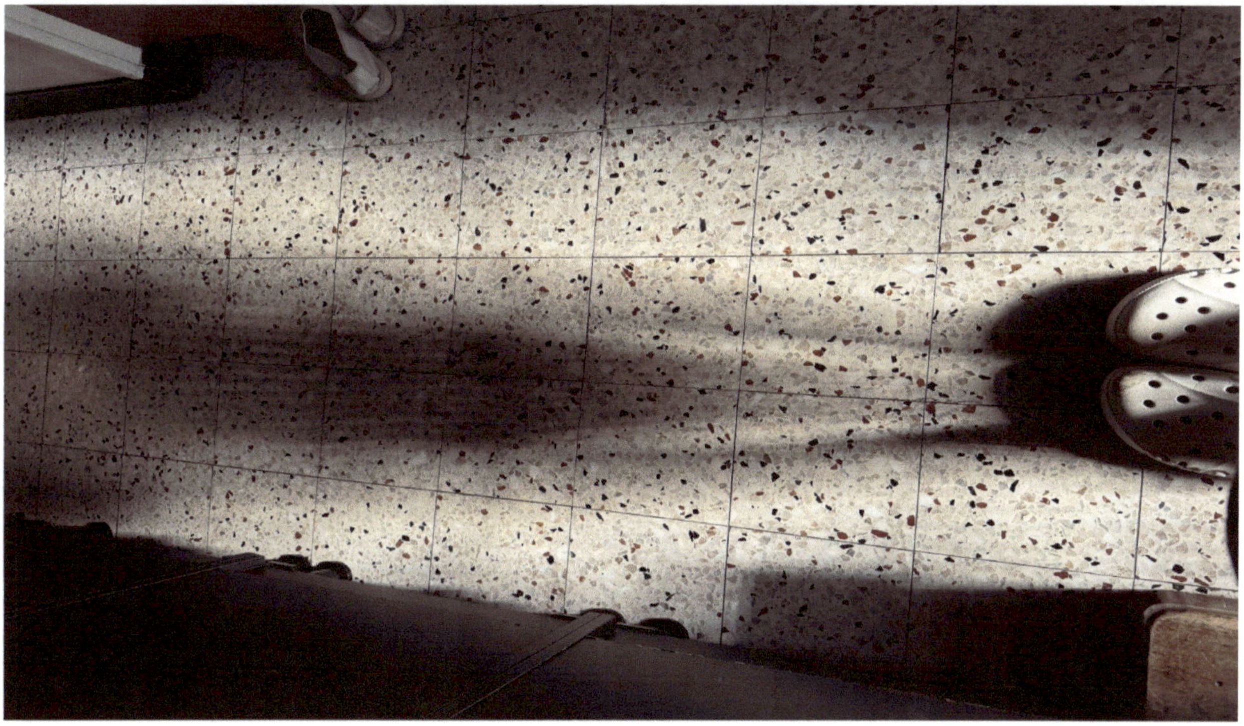

Don't forget how the light and shadow areas are formed inside a 10 cm slit, then 5 cm, with light sources with a "wavelength" of 3 cm

So what we know until now:
At atomic scale arround every group of atoms we have virtual perfect mirrors of space of FRACTALONS with differents diameters relative to the wavelenghts, so the groups of atoms will reflect a white light ,like in the photo, in the central region we see a lot of direct light reflected but if we block this light we will see the projection of the source in reflected light drom DSE of every group of atoms. In the case of DSE the wires or the bords of slits becomes cylindrical perfect mirrors with different diameter for every lenghtwave(the mirror diameter is THE DIAMETER OF THE WIRE+2X THE DIAMETER OF THE LENGHTWAVE)

This is how a point light source is reflected by a group of atoms in a mirror for different wavelengths, each pixel in our eye is formed from the sum of the lights coming from each group of atoms, direct and reflected 1.2.3.4... times, each point in an illuminated landscape sends us an image like this, but in different directions depending on the organization of the groups of atoms, the FINAL IMAGE will contain a proportion of direct central light and a proportion of reflected light exactly as in DSE

IN EVERY REFLECTED LIGHT FROM ATOMS WE ALREADY HAVE THE SPHERICAL ABERATIONS, the direct and the reflected light at differentes angles depending of the wavelenght and diameter of the virtual spherical mirrors around of groups of atoms depending of the wavelenght. The spherical aberations are not created only by reflexions on lens or mirror, they are already in the light

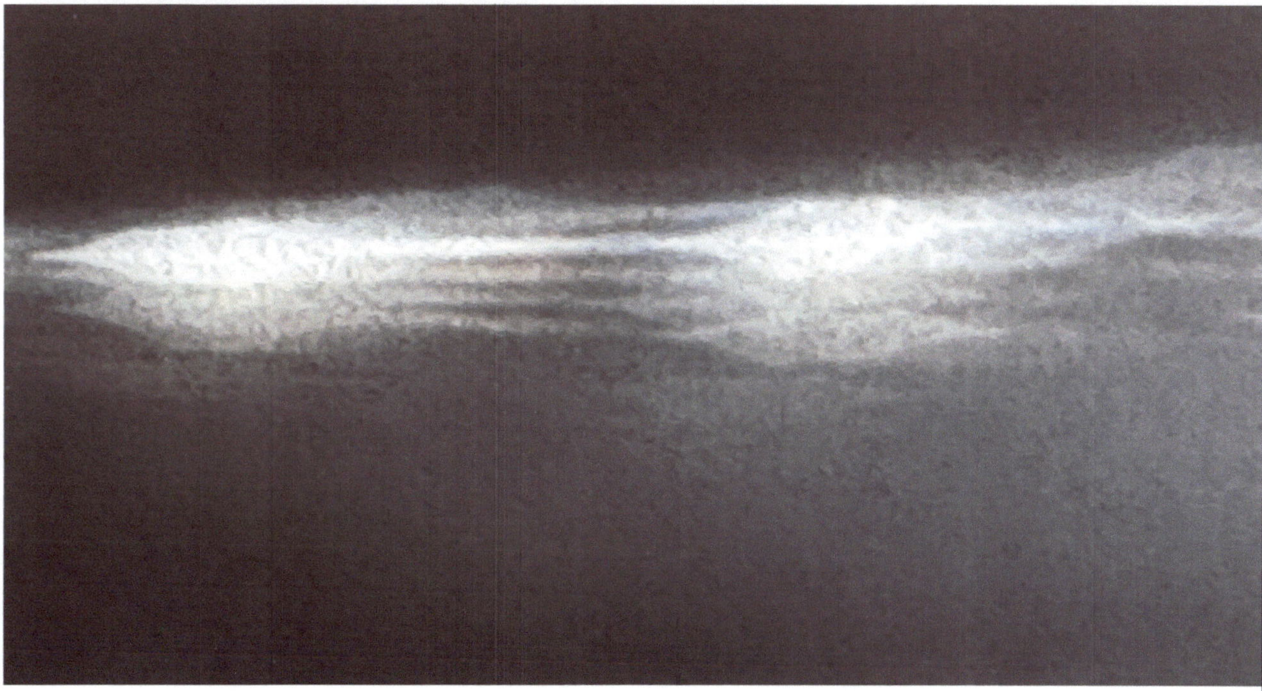

Observer view concept to understand DSE,if we block the direct light we continue to receive light from the blocked source atoms, it's just that it is the light reflected in the peripheral atoms of the mirror, of course when we zoom in we see that the wavelengths are already separated

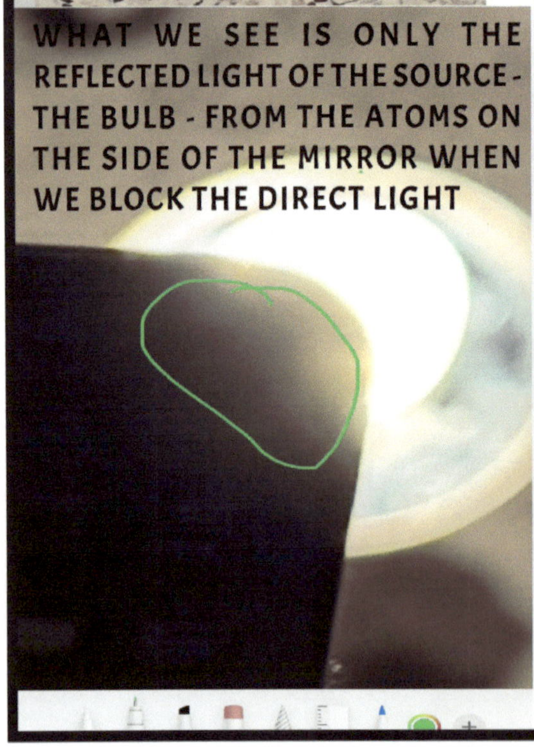

From each group of atoms in the source, direct light is no longer reflected(the red circle), we receive information only through the light reflected by other atoms which will create a virtual image in the place where the emitting atoms are located but distorted depending on the emitted and reflected wavelengths

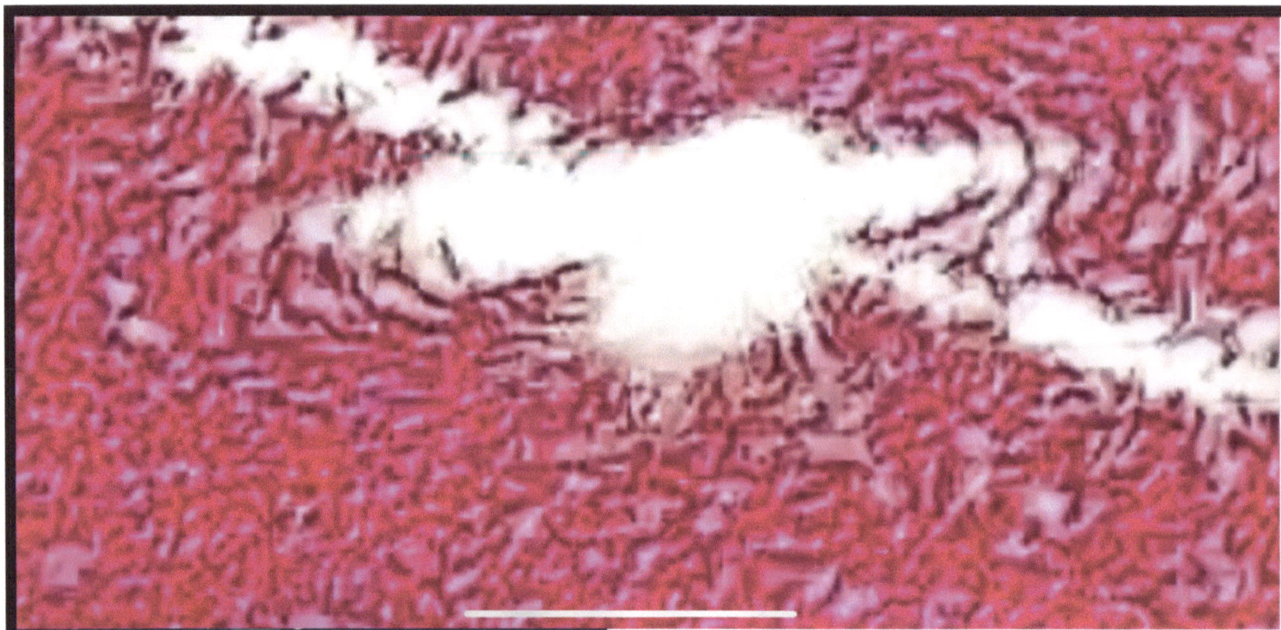

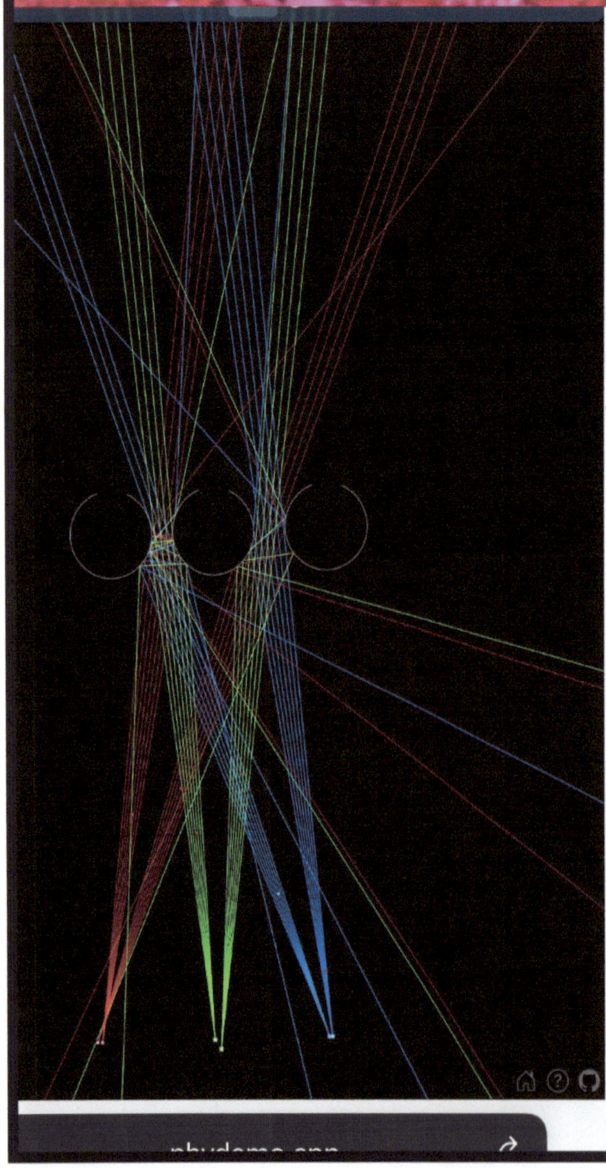

Observer - at each point on the line of reflections (I refuse to call them interferences because they are not interferences, I will present you an image with true interference at the end when the wavelength changes after reflections) Photography with Iphone 5x with maximum zoom of the two wires in the DSE from a distance of 2m

In this example, the green observer will not see any light from the source on the far midline, either direct or reflected, but he can see some light reflected from a source on the midline close to the mirrors.This light will create our virtual image on TVremote, and real image in DSE on the wall

Reality
Reality
in a
wavelengt
of 3cm
Annuler
Annoter
OK
Group of atoms in
3cm wavelength
THE TREE
Image of
a central
observer
Annuler
Annoter
OK
ZOOMED
Left observer
Right observer

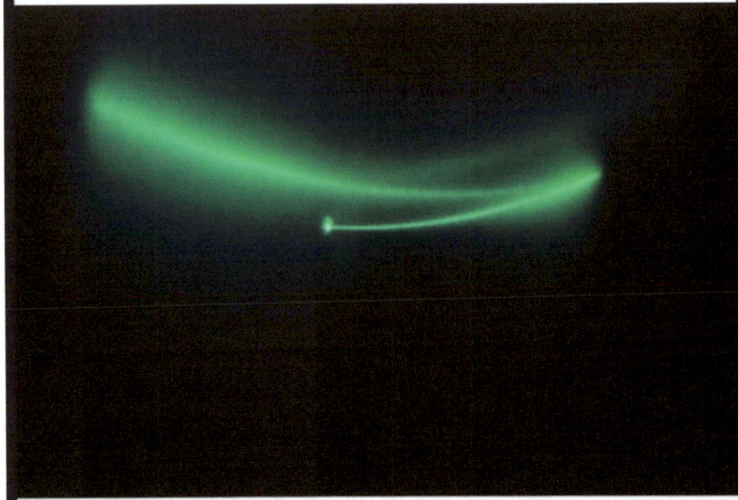

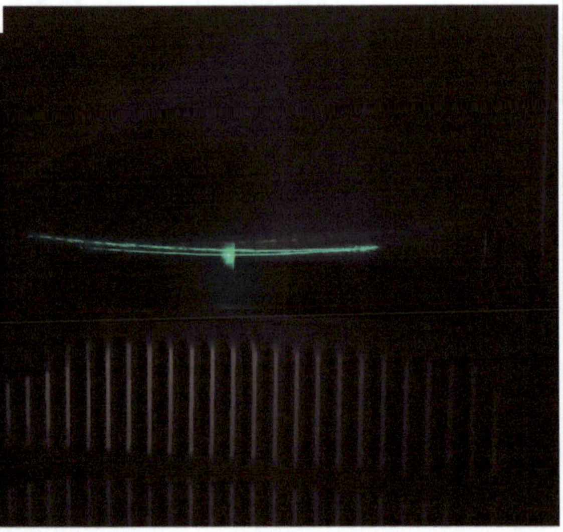

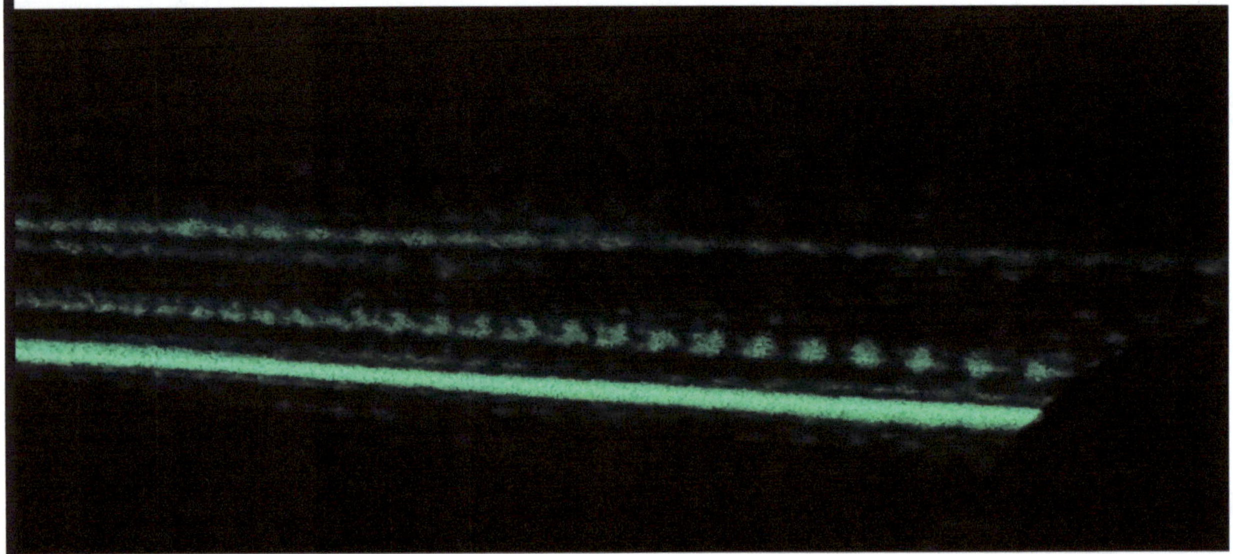

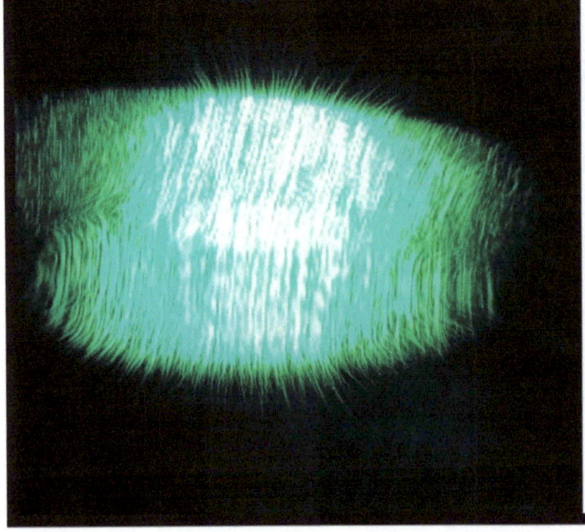

Don't forget, to understand DSE we must understand that:
1.each point in the image below will be reflected on the wall according to the geometric rule of a "Z"
2.if we hold the device in our hand and twist the wires we manage to separate all these individual lines of reflection, into the basic components, i.e. points of size of the WAVELENGTH

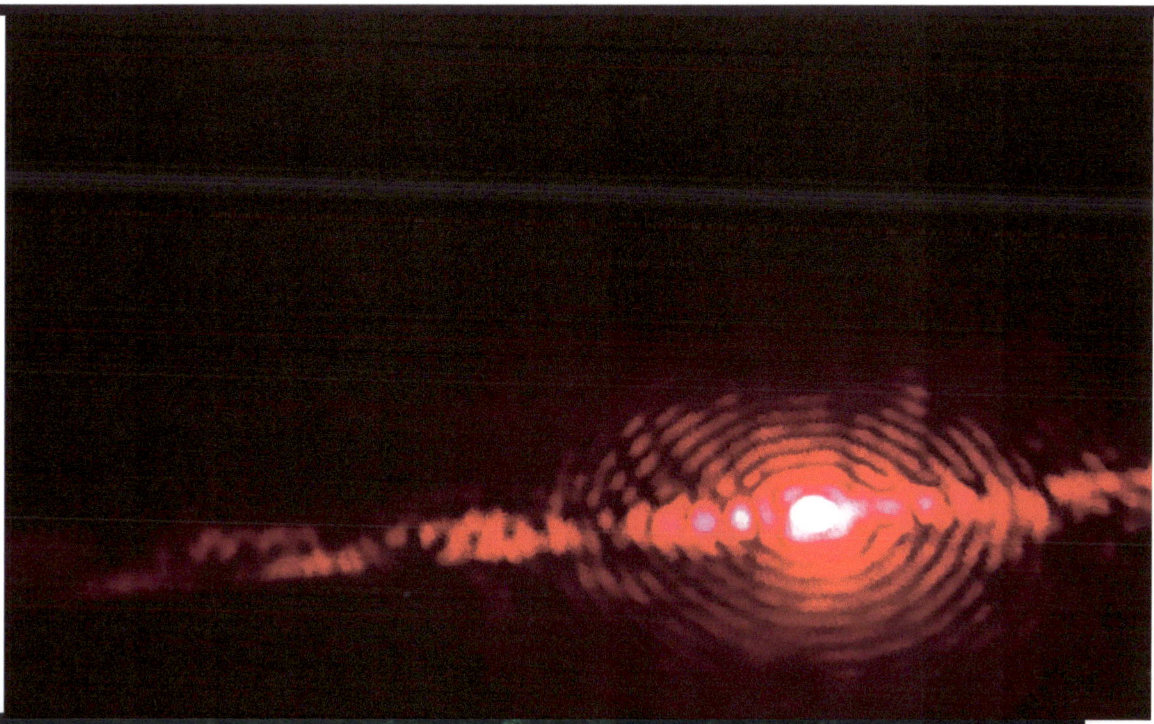

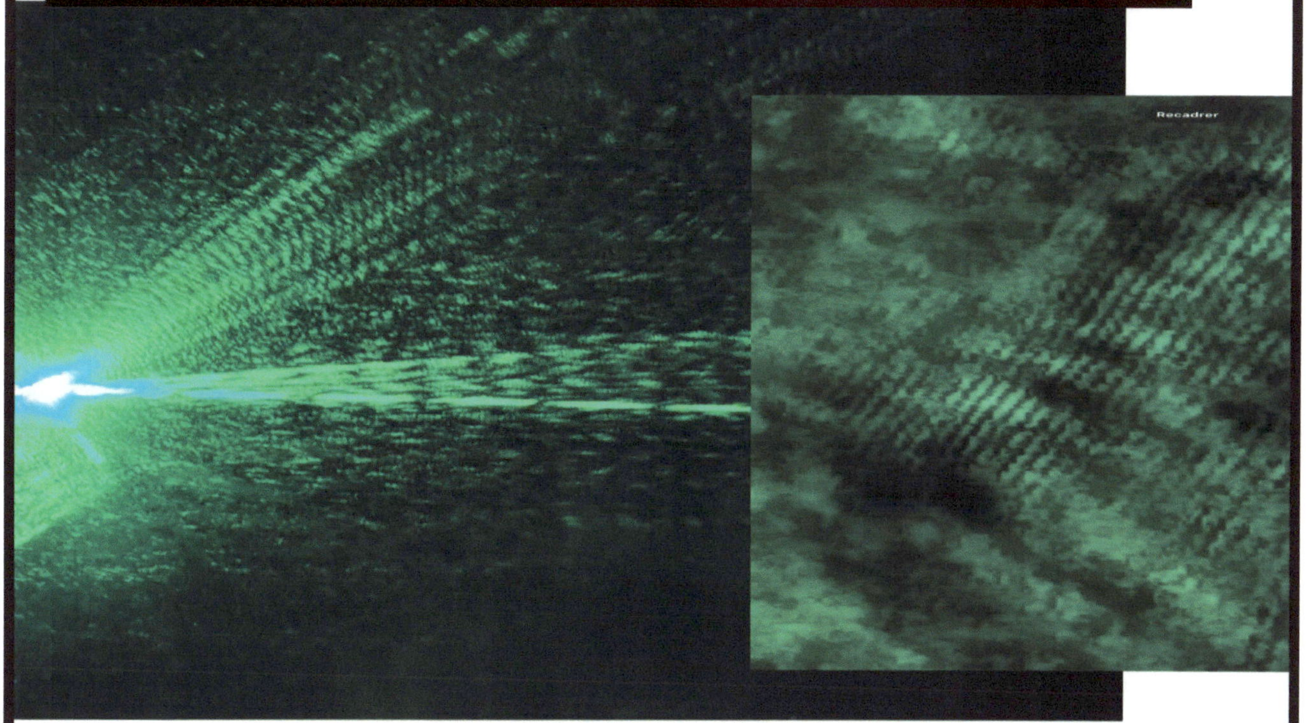

If we zoom in even harder after separating all the reflection arms of the 4 mirrors we obtain the basic image that repeats throughout the reflection, the wavelength pattern, this will ultimately form the OPTICAL ILLUSION of interference, the image built from the superposition of these tiny details reflected at different angles(the Z arms of every mirror)

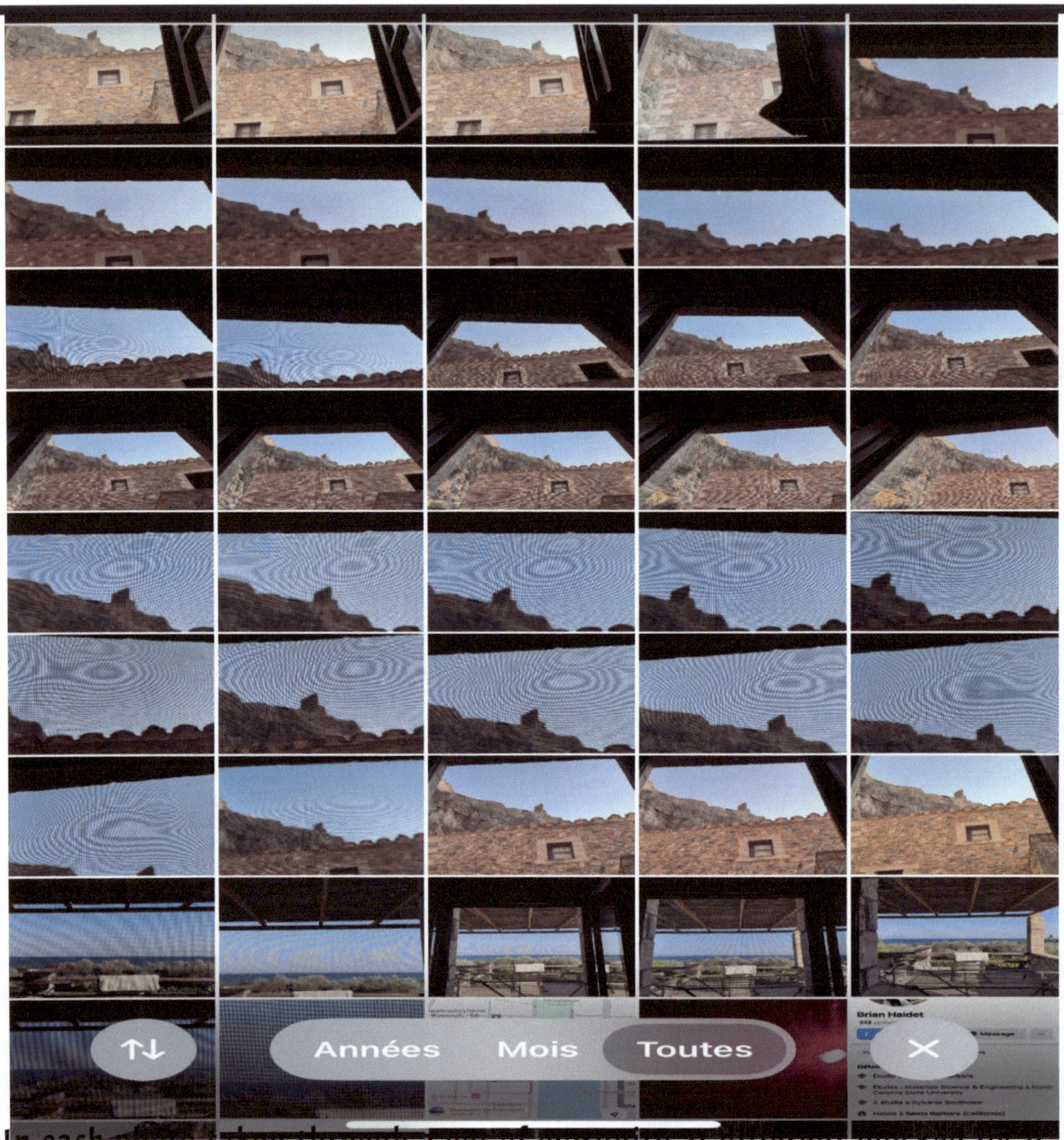

In each photo taken through a net of mosquito at different blocked focal lengths, with different lenses, even if nothing is visible in the photo and the photo seems clear due to the privileging of direct parallel light, we can find the pattern of reflected light for each color, and it does not depend on the camera, only on the light source and the illuminated scene, by zooming in and out we will find in all the photos with different lenses and completely different angles of capture, we will find the same pattern.This pattern is encripted in the pixels of image.At diferent zoom level ,different source of light and shadows.

in each photo taken through a mosquito net at different blocked focal lengths, with different lenses, even if nothing is visible in the photo and the photo seems clear due to the privileging of direct parallel light, we can find the pattern of reflected light for each color, and it does not depend on the camera, only on the light source and the illuminated scene, by zooming in and out we will find in all the photos with different lenses and completely different angles of capture, we will find the same pattern

In the same picture we recover the green from the tree reflected on each cylindrical mirror in the picture on the right and the shadows from the edges of the window at a different zoom level in the one on the left, in a picture that initially seemed quite clear

Everything that is green ,is the reflection of the olives on the mosquito net, everything that appears blue on a black background in the "interference images" is the reflection of the sky and the sea

Reflection on a stainless steel cooking pan at different focal lengths. Parallel central light, and the second picture just the light reflected in the periphery exactly like in DSE at another focal length (like in stereograms, as if you were looking through and seeing the virtual image in the mirror)

Why can we see in a 0.07mm cylindrical mirror the image of a laser with 100-150 and more wavelength striations?

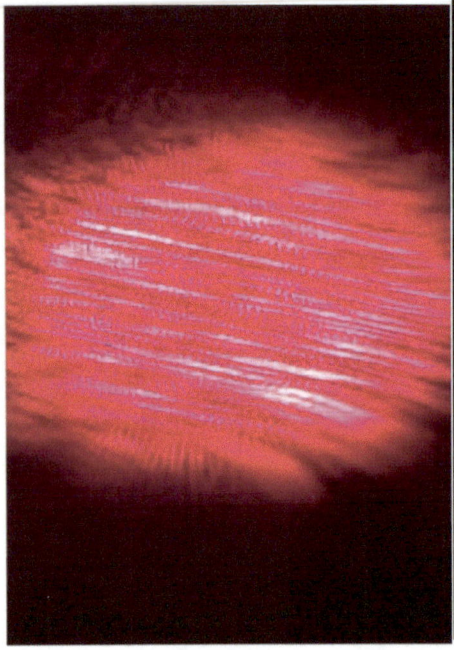

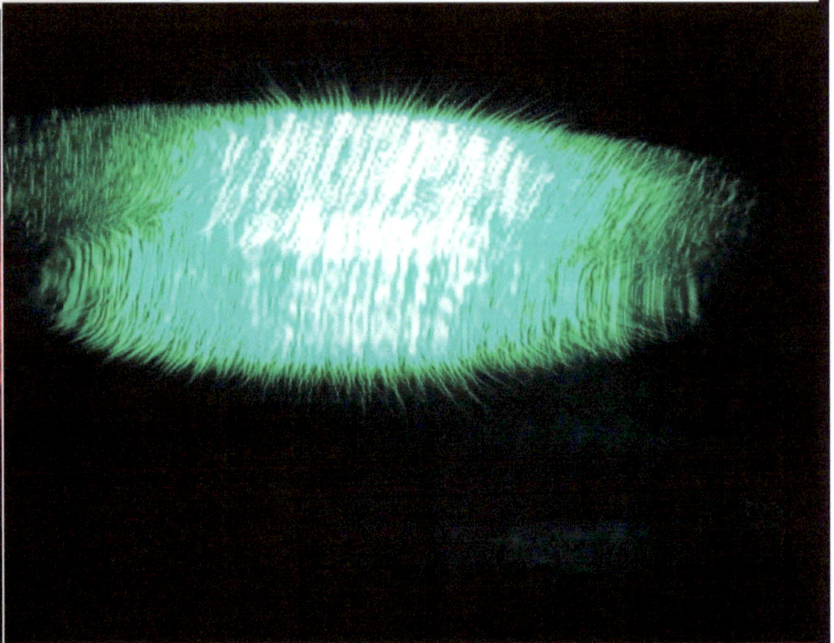

Because on the surface of the wire each quantum of light creates its own mirror, the mirror is virtual, not material, it is made up of fractalons from the vacuum, and their size is the size of the Higgs's boson, the minimum size in the universe (Planck resolution)

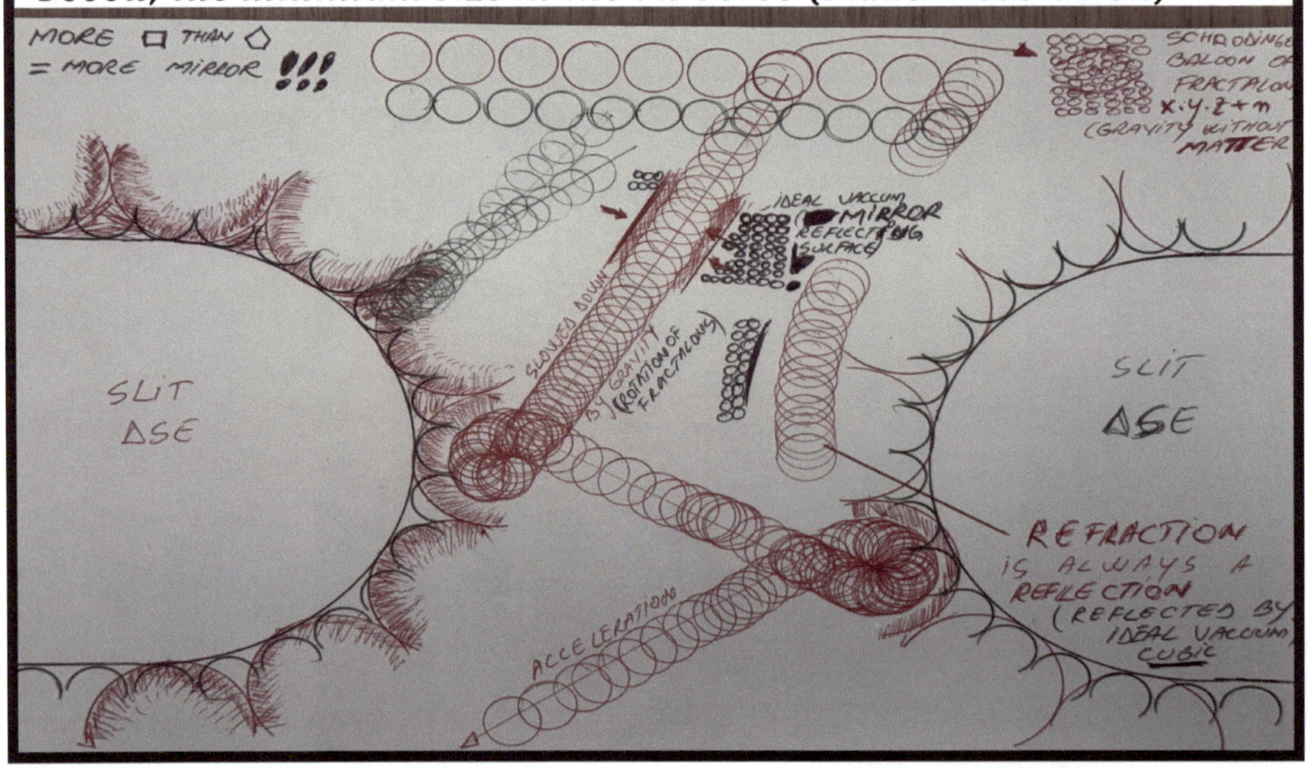

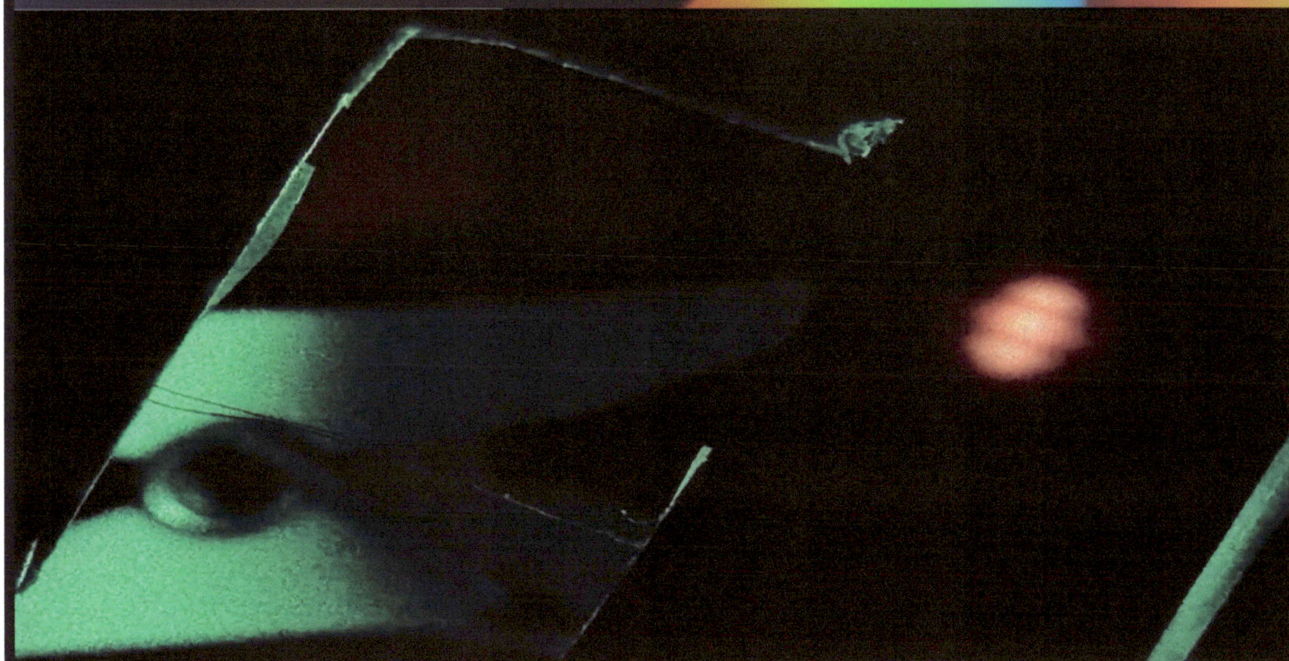

Real interference, light perception(in fact, color is just a perception of our eyes and brain about of a quantity of energy) changes wavelength, green laser on credit card plastic, or two different reflections from a CD seen separately or together

So if you hold the wires or slits from the DSE in your hand and twist them a little and move them carefully, you will notice with the naked eye that on the wall you have the classic projection of several cylindrical mirrors that move independently of each other. And with the little tips that I have given you, you can also answer the question of the century.

Through which slit did the photon pass?

In fact on which mirror was reflected the last time.

Now comes the hardest part:
1.to understand how reflection on cylindrical mirrors occurs
2.how the reflection of different wavelenghts occurs at the atomic level and MUST IMPORTANT at the FRACTALONS level
3.What the frequency of light is
4.Why light and gravity have the same speed and what is the difference and the similarity between light and gravity.
Unfortunately I am not at all good as teacher or as a youtuber, so I will appeal enormously to you inteligence.I will present you visually the consequences to you and you try to understand them and assemble the puzzle.
To succed, i will give you 3 helpful elements.

1.How to average several elements without doing the calculations based on the fact that nature does better calculations than us, for example when we trow a stone the best way to find its trajectory is to actually throw it.
2.How to find mentally the intersection of two equations in nature(like Schrodinger equations)without doing the calculations, we imagine where they intersect and we kind of know in 3D XYZ what the solutions are
3.and the last one and more IMPORTANT is a small synthesis of the Fractalon Unified Theory presented in detail in the volumes FRACTALON 1,2,3 on Amazon and videos of my website fractalon.fr, i will put here a "resume" because i was banned to present my theory in Youtube, Wikipedia,Quora by this wok generations of liars physicists and apotres of them.

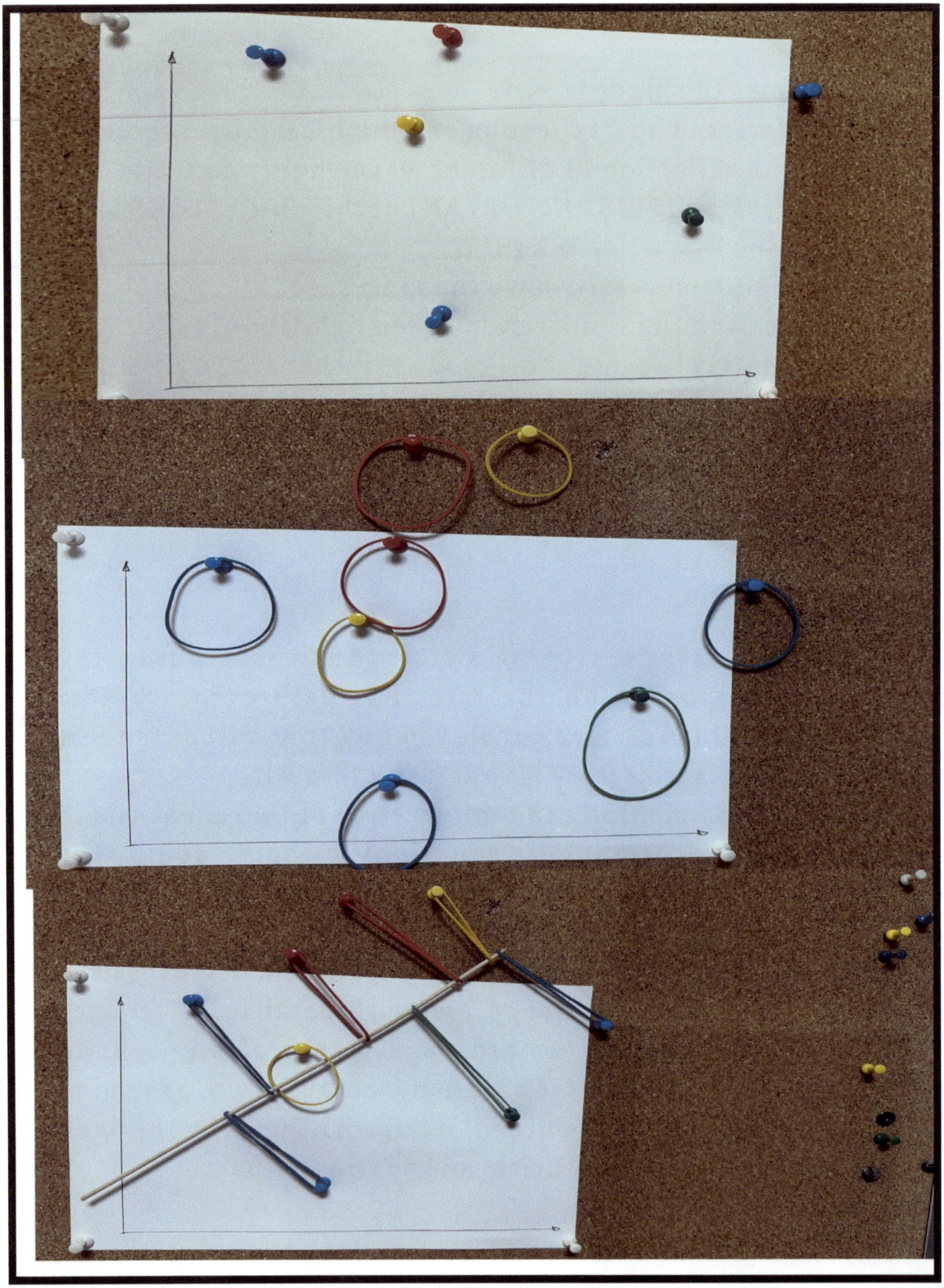

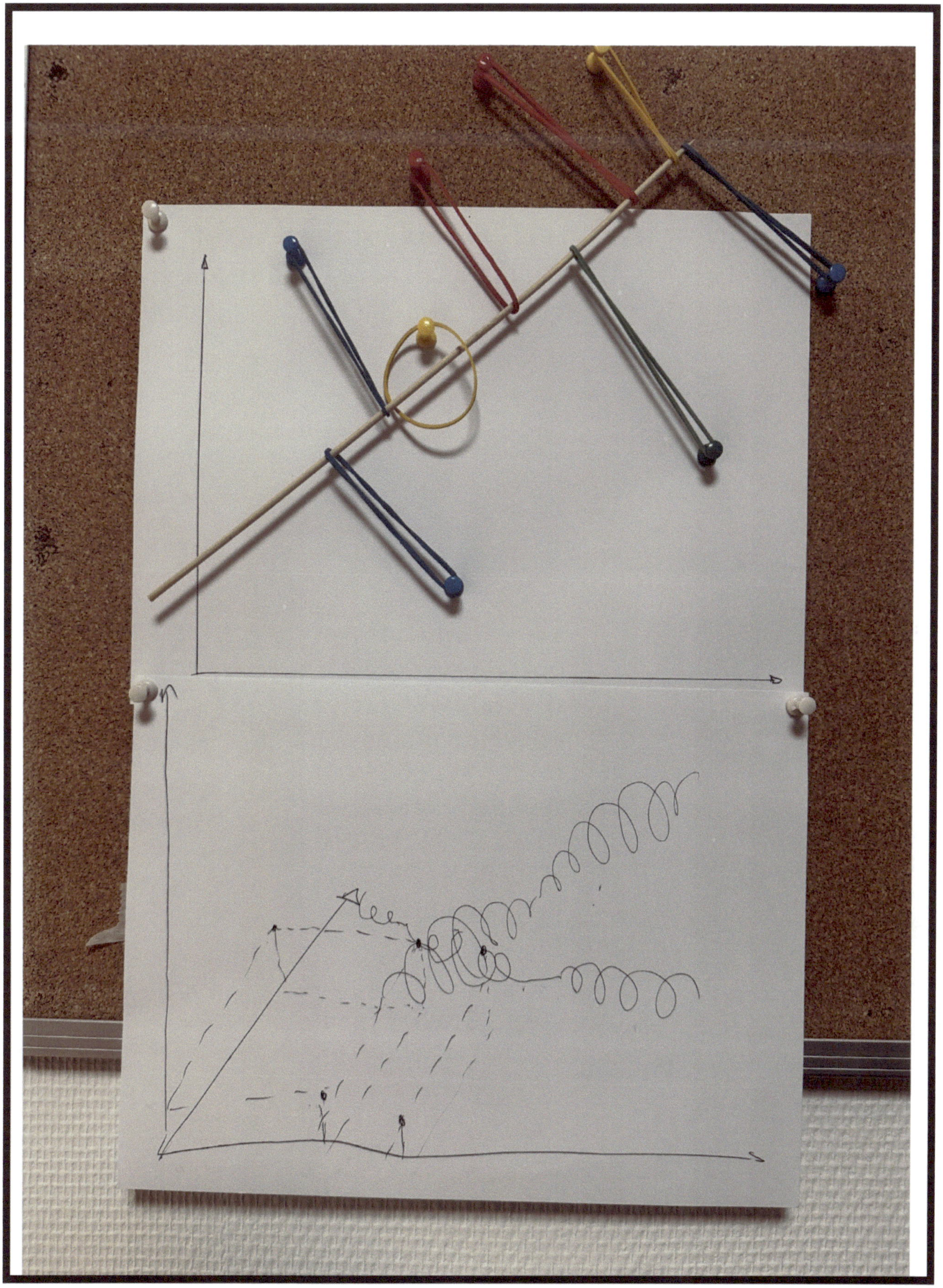

Now ideal is to read FRACTALON 1,2,3 and to look at videos of fractalon.fr or if you are very inteligent just go to page 145 to see a little condensed synthesis from my banned Quora posts and look at this logical order of photos to understand what we see if we zooming at lowest level in our univers, at the level of fractalons, electrons and atoms.After come back here to continued.

https://youtu.be/XVOmlQfT__k?si=jF-Y3cbzJ1E623cp

Why!? Because it will distort space and automatically time, into concentric spheres, the equatorial contact will be lost more and more towards the additional boson.

https://youtube.com/shorts/dsJrCt8YtvI?si=J1hKTnAoZ7-VAhMw

an additional fractalon in the cubic structure will receive a double 8 movement THE ELECTRON spin,and create two waves: GRAVITY AND ELECTROMAGNETISM -fractalon.fr videos

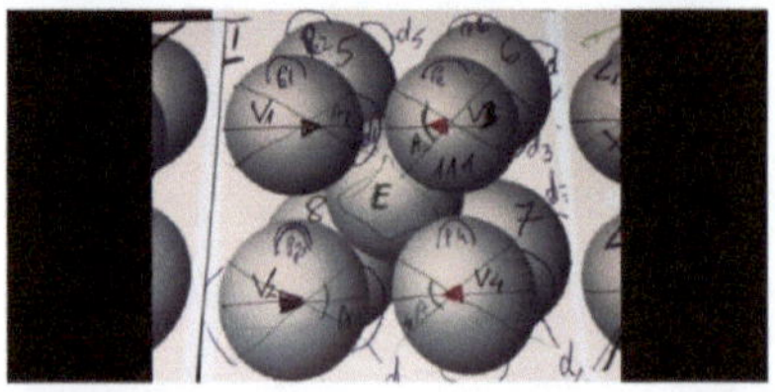

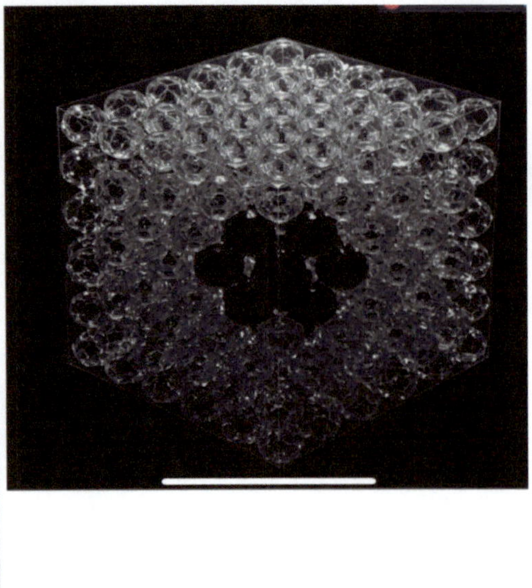

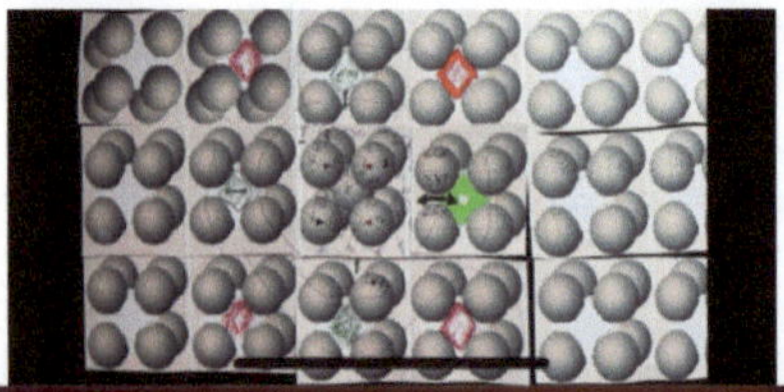

Gravity and E-M wave of precessions

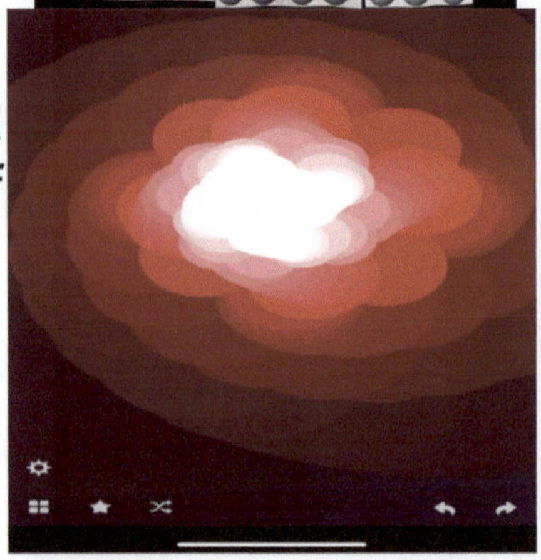

IN THE LACK OF SPACE, electrons crystallize into a PROTON

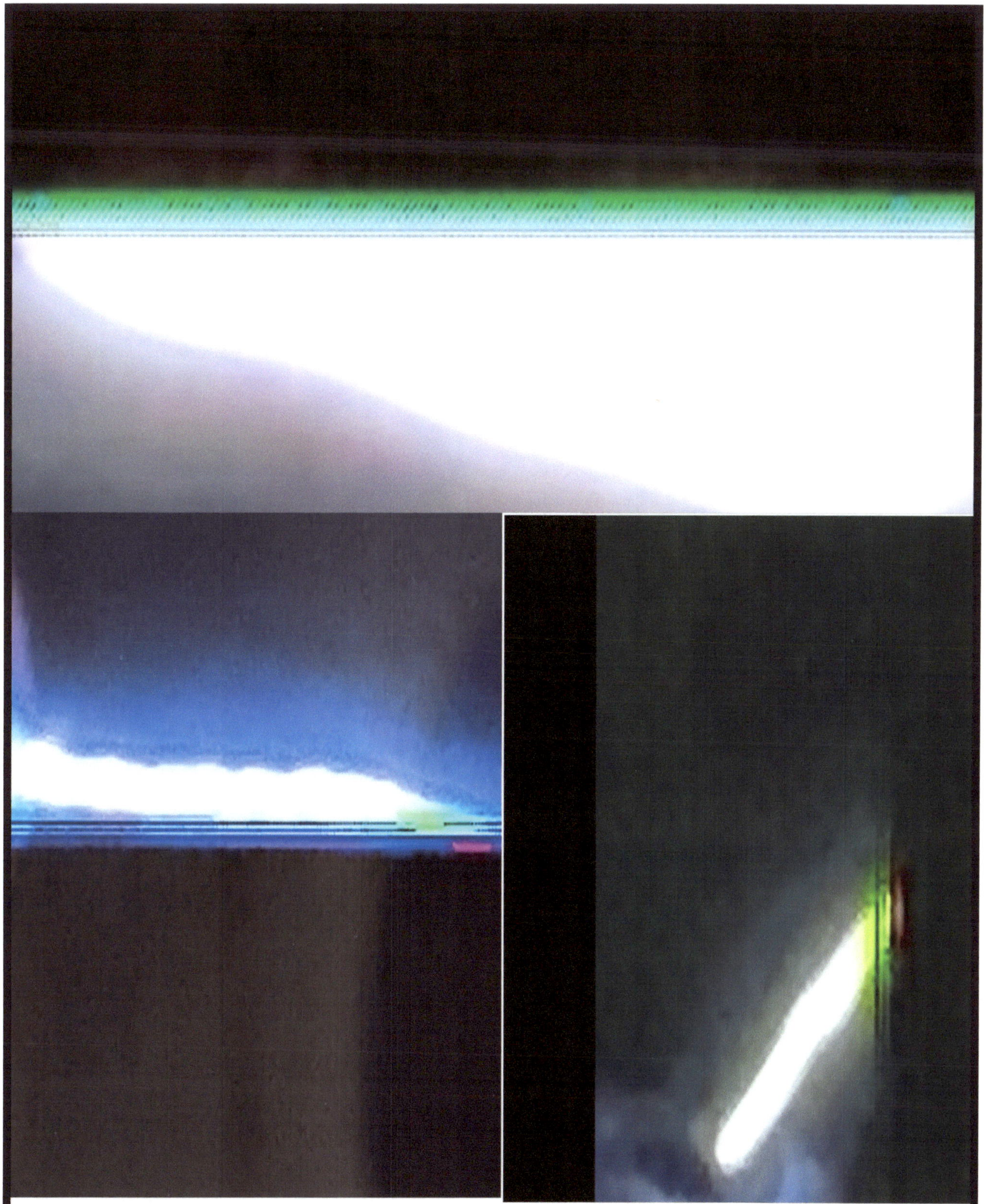

"Spectroscopy" with Iphone 5x camera of lighting strike , order of emiting light, order of exciting in suplementarry electrons of the layers of space and order of decay of electrons precessions pression in wavelenght

Do you REMEMBER? :1.DECOMPOSING REFLEXION image on cylindrical mirrors DSE like.
EVERY IMAGE created by cylindrical or spherical mirrors(and pratically everything we see with our eyes)can be decomposed into
1.1 DIRECT LIGHT(in which we have direct light from the source and reflected light from the source=reflected in the superficial layers of the vacuum upon contact with the source, after its creation and will give us a different pattern of the light source for each wavelenght, a kind of ocean of standing waves at resonance,a bucket of water with square walls that is hit a certain frequency by the bottom.
Some images to understand this:

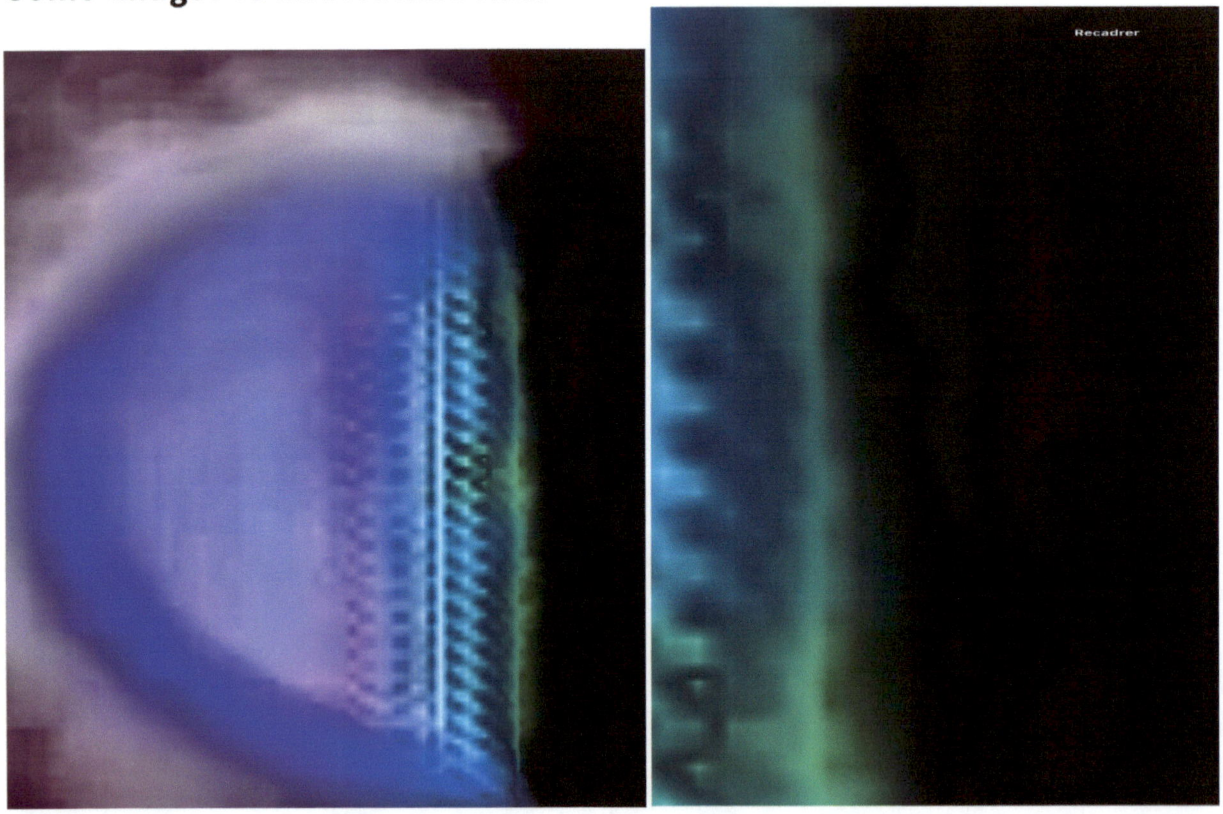

Photos zoom5x with an Iphone of a lightning strike in the Wimshurst machine.
Each wavelength is emitted in a certain order and with a certain pattern depending on the order of reaching resonance in the RAINBOW-WOBNIAR fractalon space. The electron detection resolution (much superior to the resolution of visible wavelengths) of an iPhone sensor allows the visualization of some of these patterns.

We get a little idea of what our DSE light source for the green wavelength, with an almost perfect geometry, actually looks like

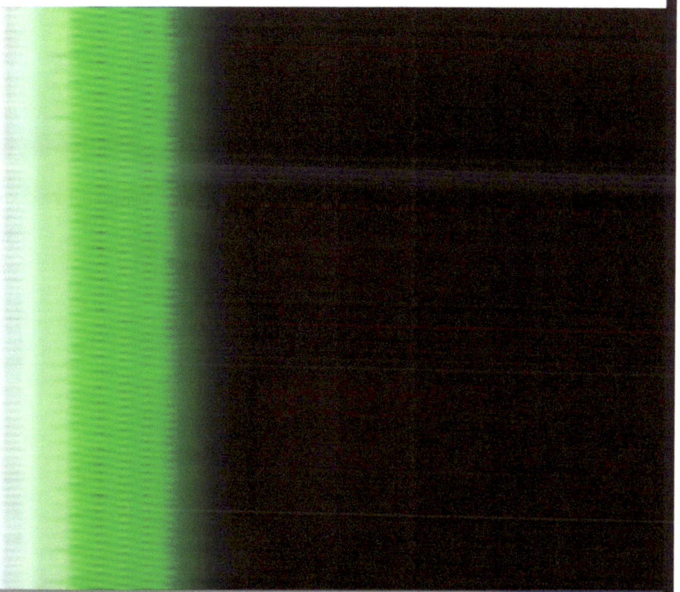

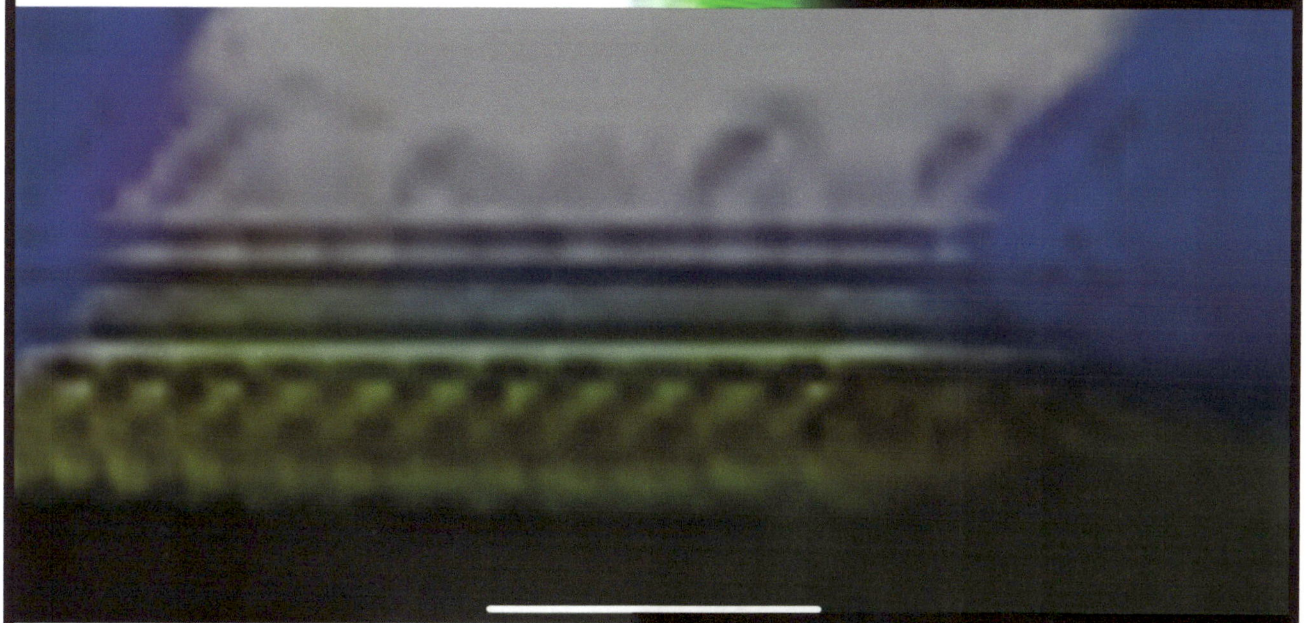

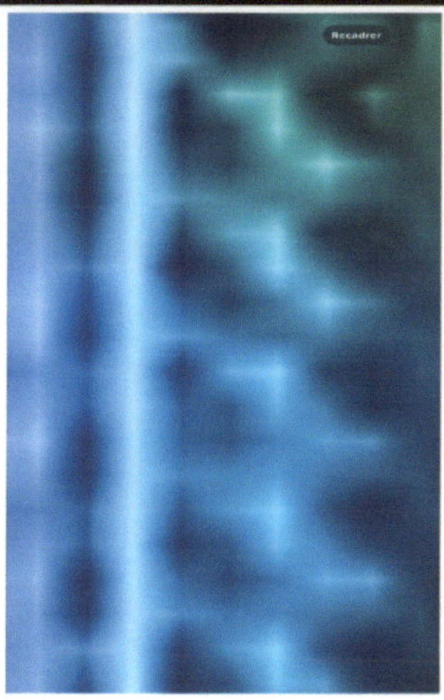

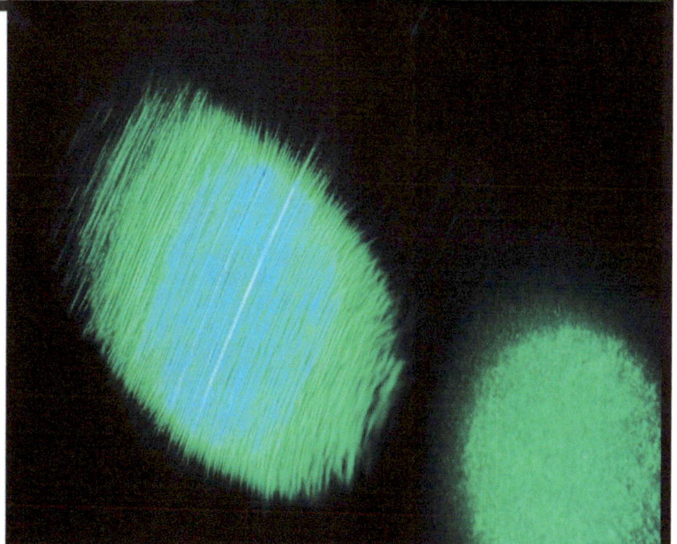

Layers of pressions=layers of differentes precessions=$1/n^2-1/(n+1)^2$

Photo with iphone 5x of a lighting strike and second photo few miliseconds after the lighting strike, to see how ether is organized in bubbles of differents pressions.

One atom or one group of atoms cannot emiting in same time two different length wave ,it's all about pression so you reflect or emit depending of the precession or rotation of neighbors (Fractalon part 3) depending of gravity (rotation)and electromagnetism(precession) but precessions gives us a region (a ballon ,a Scrodinger 3d balloon like in the video of Steve Mould with resonance of a ballon) whit similars frequency of precession and a fracture limit like a invisible soap ballon(Fractalon first part explanation).

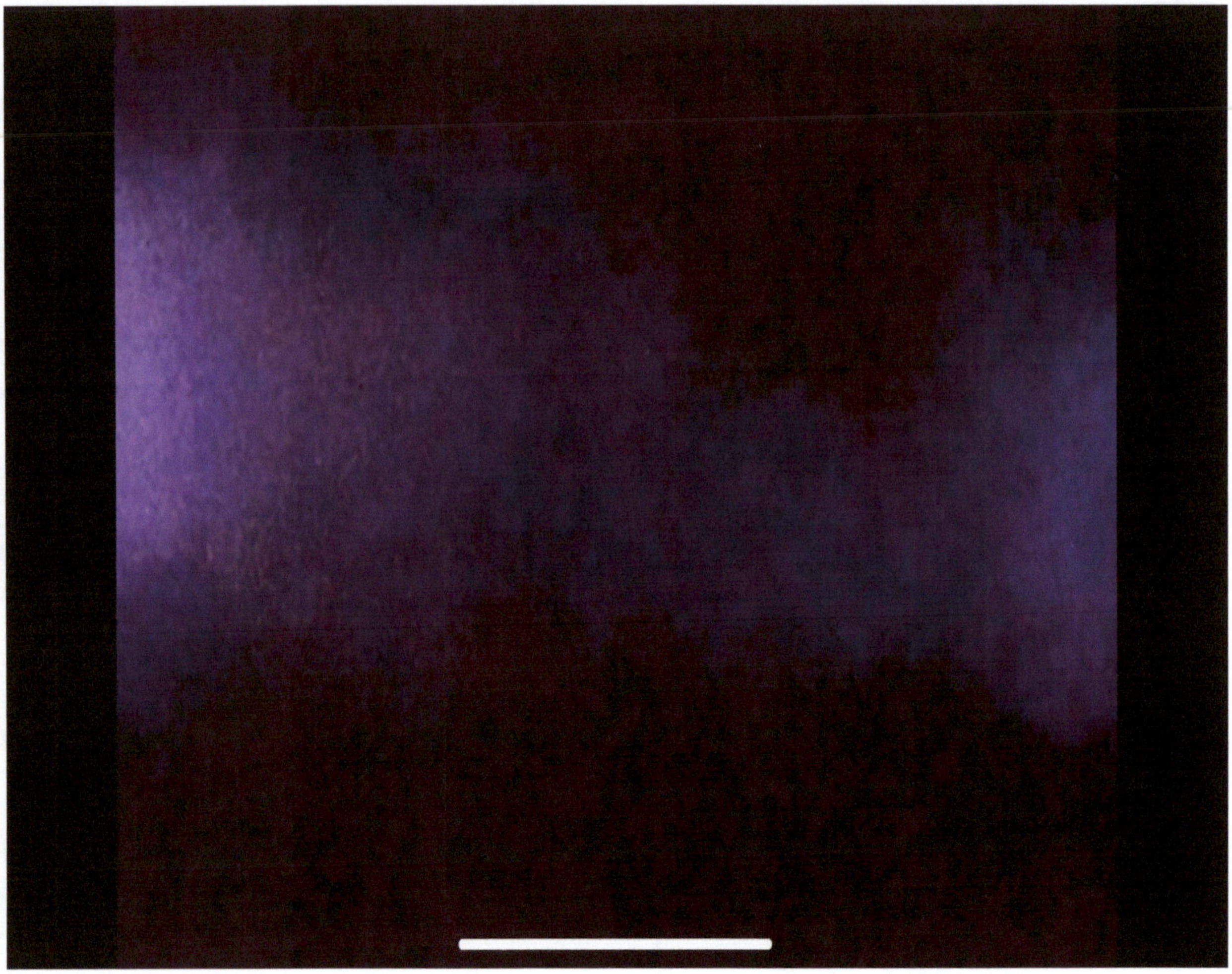

If you cover the metal spheres of Wimshurst machine with glass you obtaine a continus red and after just another unique wavelenght of violet beacause the organisation of space is different in atoms and arround them(the medium , the air don't change but the modality of electrons to travel changed, they travel incoded in violet light and and will decharge only certains levels of electrons of the fractal metal of the machine, you can decharge Wimshurts machine in succesive etapes (see my videos), every lighting is another peripheric layer of GRAVITATIONAL- EM layer of electrons decharged

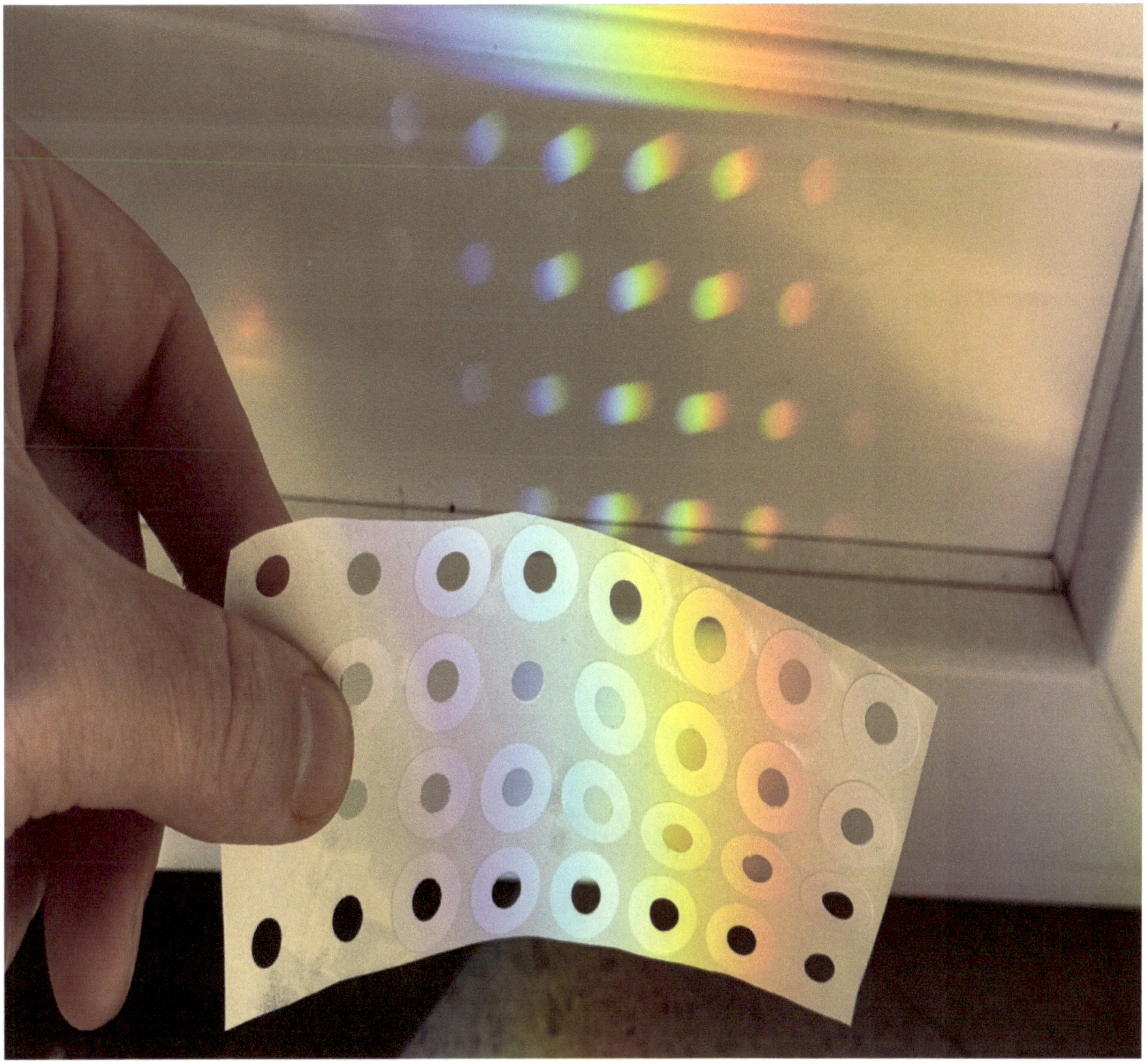

If we have two different light sources, as in the reflection of a CD, and we block one of them, we see that electronic sensors, like the human eye, detect energy per unit time and not light. A strong red or green light will be interpreted by the eye as white. And different combinations, like different colors (theory of color combination)all in fonction of total energy per time unit.

3.1 WAVES, what is the connection betwen gravity and light...and sound wave and all the waves...? Why gravity is the mother of all the waves and why the space is the father of gravity and time, ROTATION and TRANSLATION(PRECESSIONS of FRACTALONS)

If you ask a physicist in 2025 what he sees in this picture, he will tell you that he sees a failed interference picture in which interference between sea waves and the waves of a boat is seen, and that it proves that waves are waves and form interference.

If you ask me what I see in this picture (I chose another picture because I never want to be in the same group as today's physicists, with a few exceptions in my family -FR, I will never forgive any of them)

me:

In this picture we see a ratio between the energy transported by the gravitational wave (in our case at the boundary between an air column and a water column) and the energy transported by the ratio between the deformation of space created by a group of atoms and their translation energy in relation to the energy of the gravitational wave around them, so in Romanian, an oscilloscope of gravitational waves, and their interference with matter (translations and rotations). We also see the reflection of similar series of gravitational waves that some call the wavelength of electromagnetic waves upon contact with matter, so an oscilloscope of different lights and which helps to measure the speed of the gravitational wave. So the only wave that is seen in this image is the gravitational wave. In this image, the speed of sound that depends on the speed of gravity is not seen either, and the speed of gravity is not seen either, it is only seen that certain oscillations of gravity are reflected each under a different angle. And to be even more correct, they are not reflected, they are slowed down, deflected and turned back.

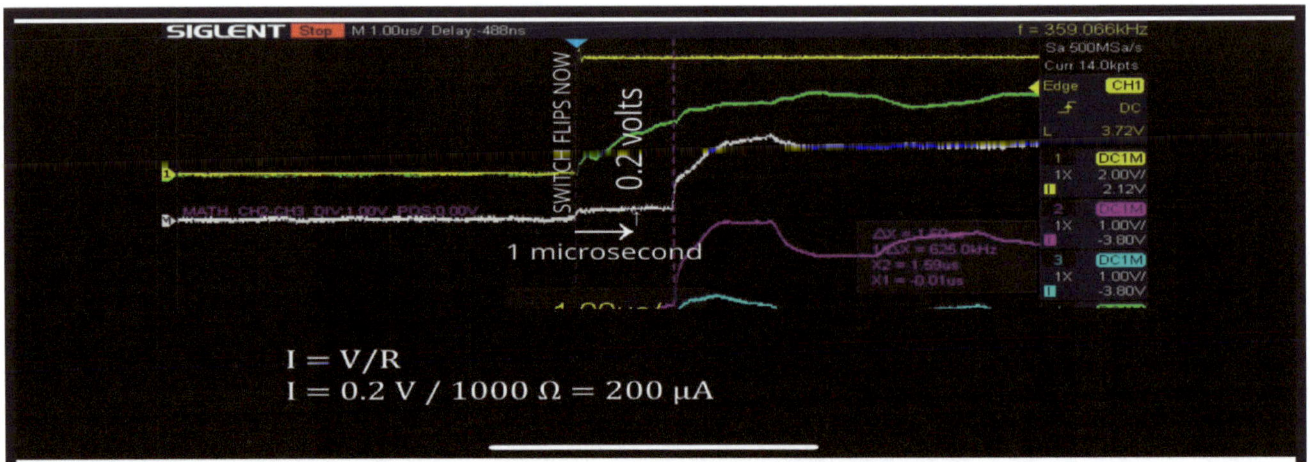

...25 hz wave sound, 50 Hertz, 60 hertz electric current,650 nm wavelenght, etc, are just mesure of something encoded in original wave, 25Hz encoded in the wave sound,50 Hz electric current oscilation encoded in the light wave (but in fact every light frequency and lenghtwave are encoded in something else, and every sound wave are encoded in something else.When you block the molecules to pass their energy in one direction the gravity is reflected because the mass is reflected , size matter in universe, because create deformation of the space , but in the end all the waves we see are encoded in GRAVITY.So what we see in this photo is the diferents speed of GRAVITY at different distance of atoms, different speed in differents layers of GRAVITY arround wires, and we see a little part of current are carried in let' say radio waves in layers of cubic fractalons structure, and the big part of electrons are carried in the "train-asteroid belt" of let's say green wavelenght in a intermediate structure of fractalons with lower rotation speed of fractalons, lower speed of gravity

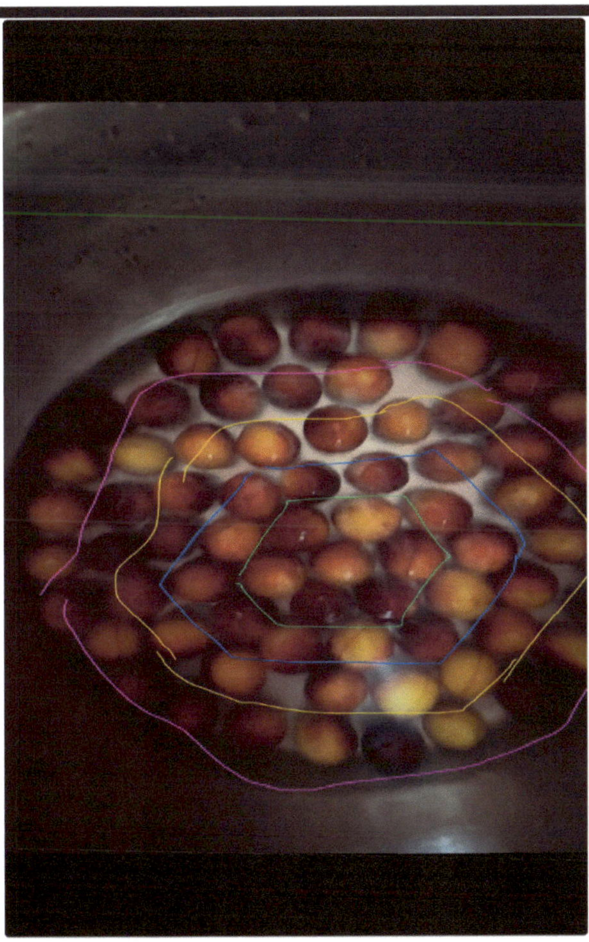

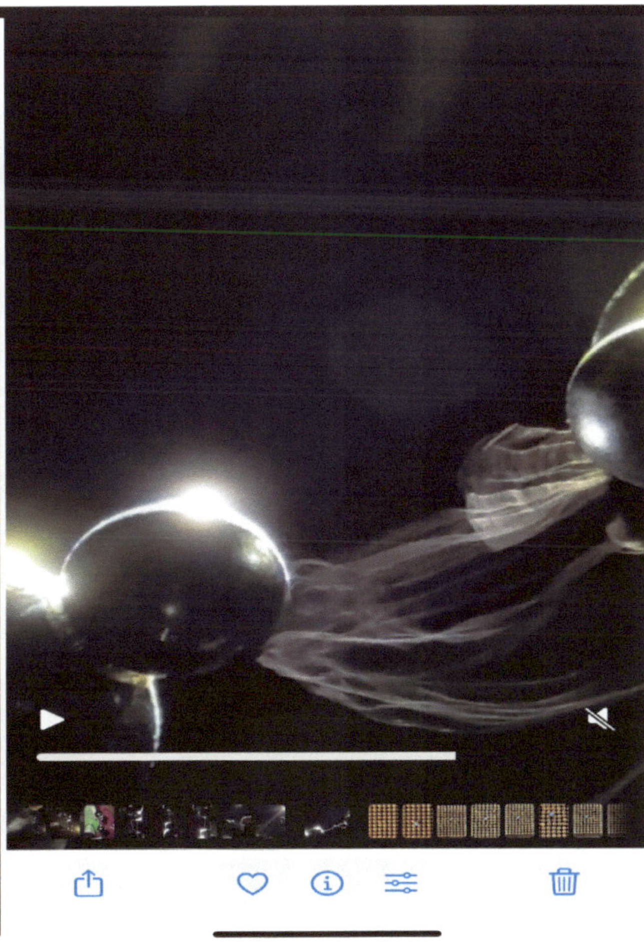

Different layers of fractalons electrons around a proton, each layer having similar precession of fractalons, each layer will exert a pressure on the inner layer, the electromagnetism is always engraved in gravity, gravitational fractures of space will be the path of lightning, and we can visualize them before they happen-video fractalon.fr

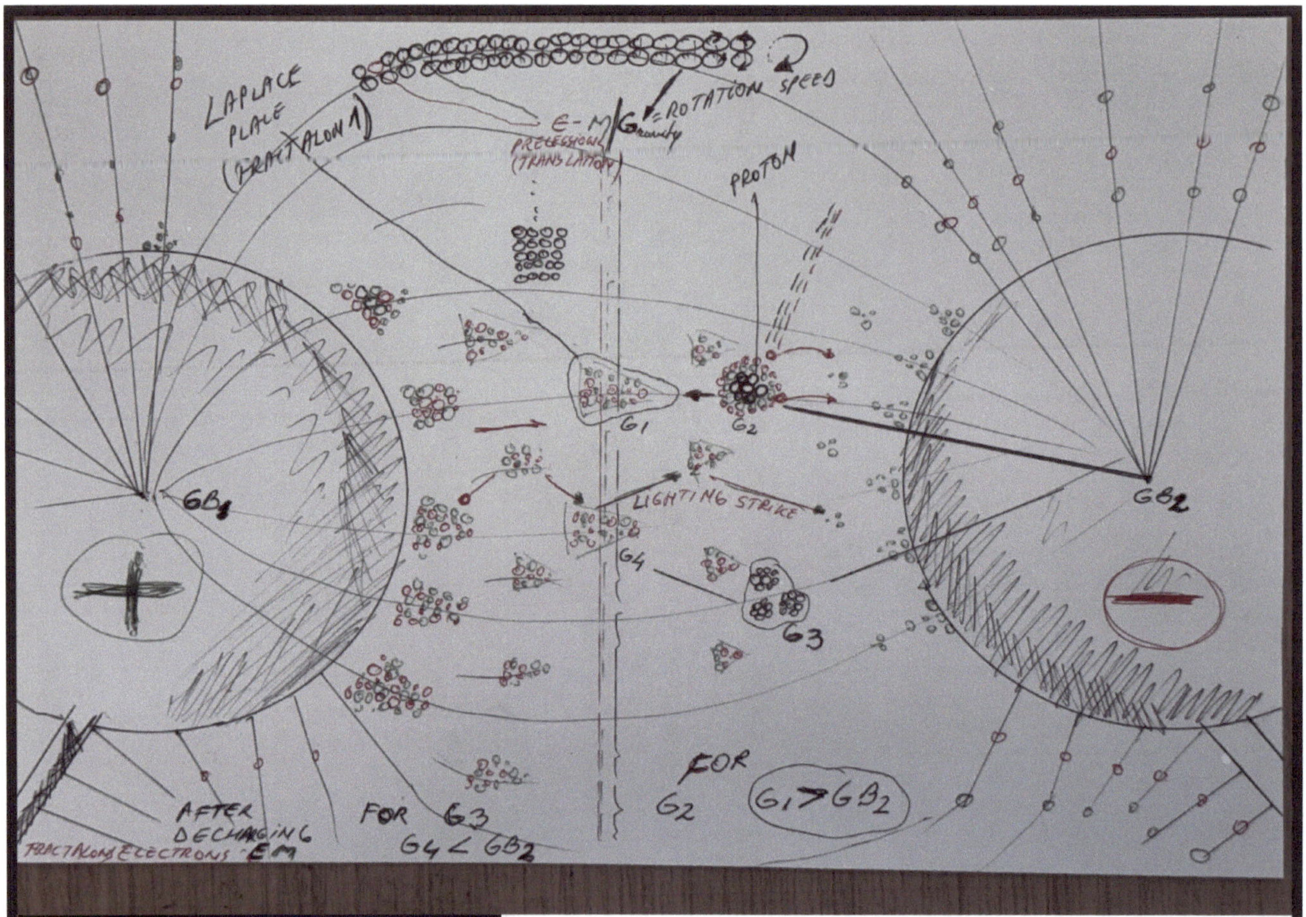

A dust particle reflected at a certain distance from the metal ball, after the electrons fractalons pressures equalize (transfer of additional electrons to the deficient metal ball), normal gravity will take control over it

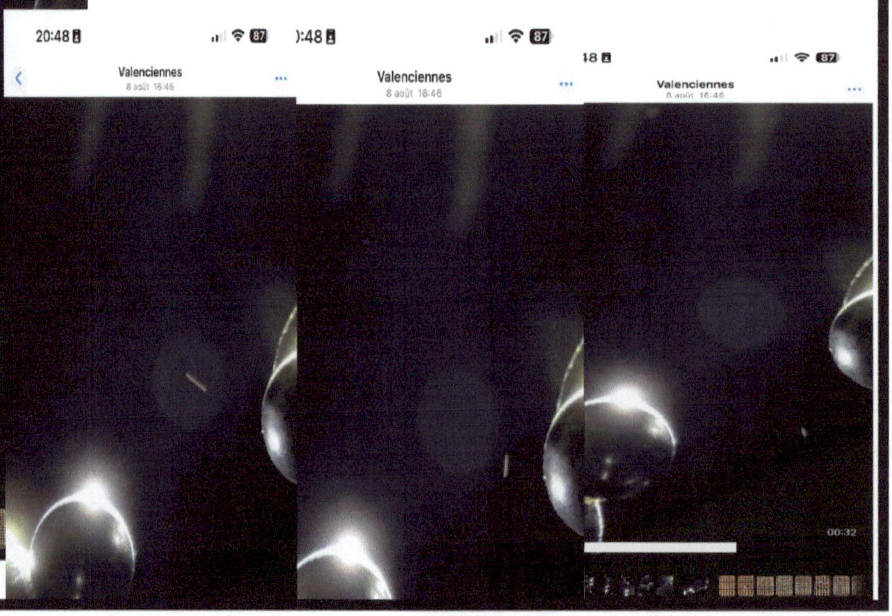

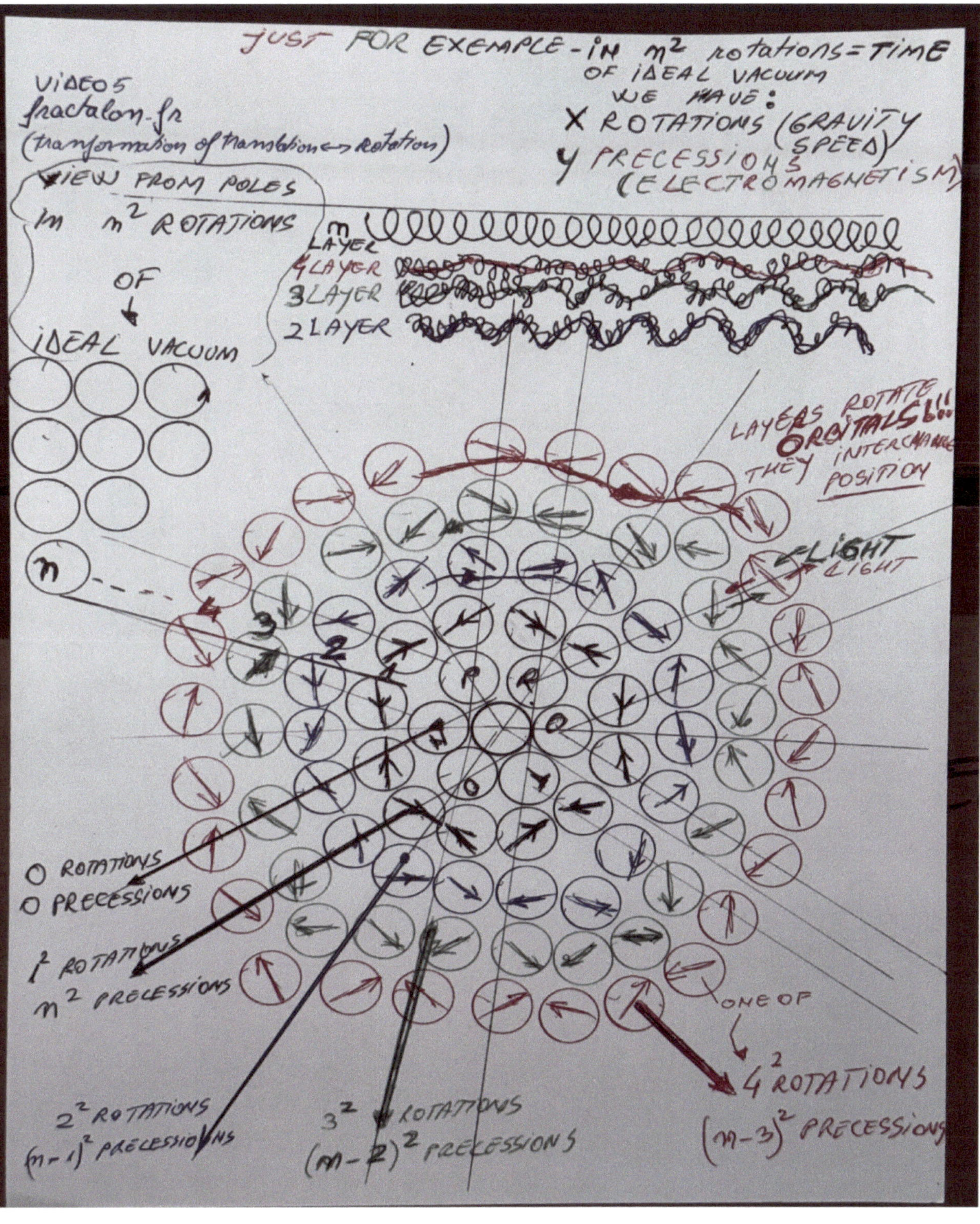

The layers of fractalons electrons around a proton and the pressure difference through beating, through precession, between different layers that will give us the electromagnetic spectral (Lyman, Pachen, etc). Each wavelength will be encoded, transformed, accelerated or slowed down, by the gravitational wave between different layers by the loss of equatorial contact (FRACTALON 1, or the videos fractalon.fr)

Ideal vacuum in the periphery, a HELIUM ATOM in the centre,and all the electrons arround it(ultra ionized, a lot of suplementary fractalons,a lot of vibrations and precessions)

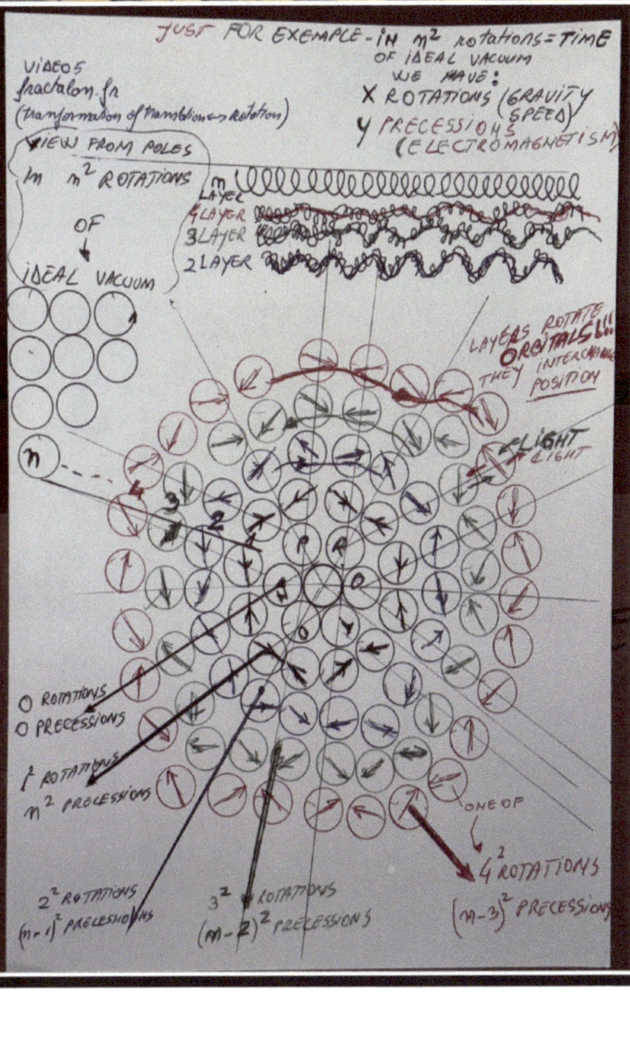

A proton with its electrons around it, in each layer, 1,2,3,4,... we will have another configuration of rotations and translations and especially precessionof electrons in double 8. Video fractalon.fr

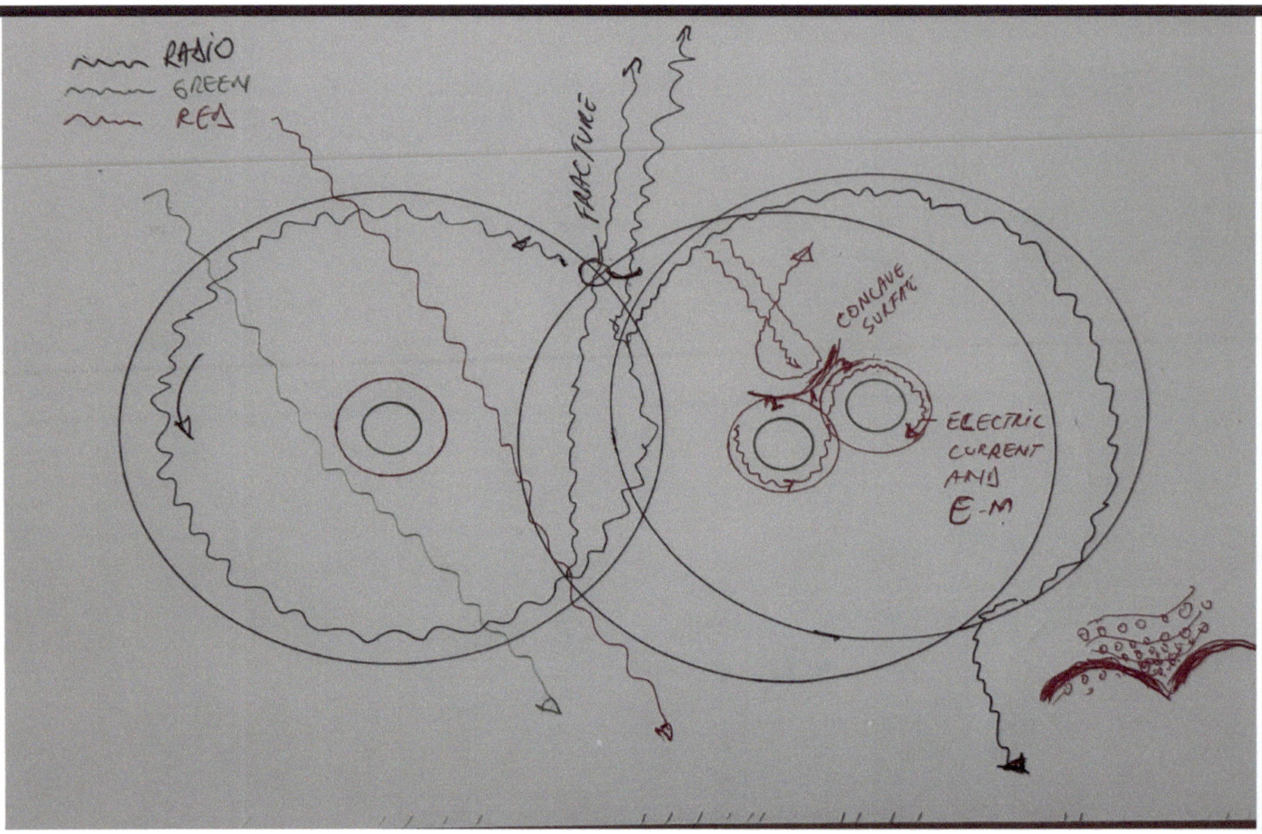
RADIO
GREEN
RED
FRACTURE
CONCAVE SURFACE
ELECTRIC CURRENT AND E-M

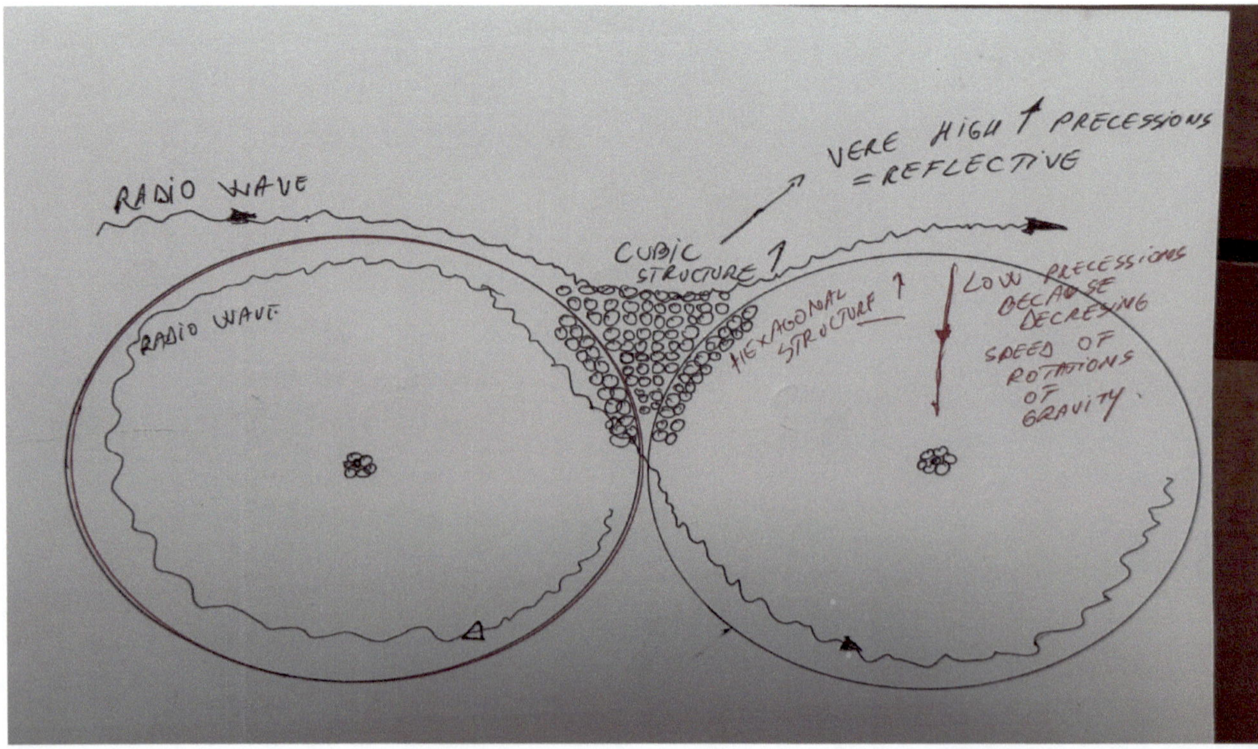
VERE HIGH ↑ PRECESSIONS
=REFLECTIVE
RADIO WAVE
CUBIC STRUCTURE ?
HEXAGONAL STRUCTURE ?
LOW PRECESSIONS BECAUSE DECRESING SPEED OF ROTATIONS OF GRAVITY.
RADIO WAVE

Schrodinger's light balloon

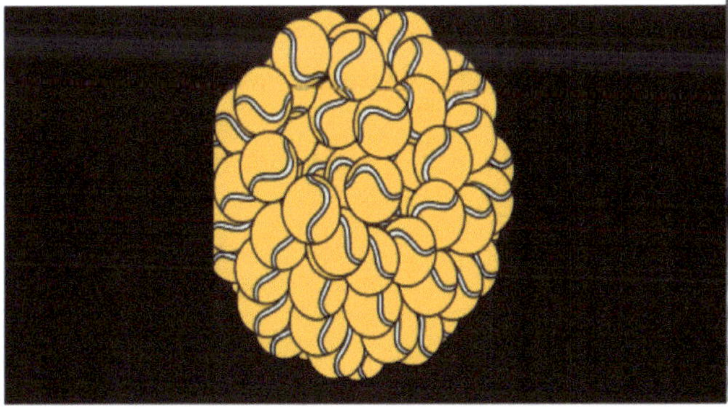

When in an ideal vacuum volume of XYZ fractalons we add n fractalons (1.2.3 etc., for example for the red color there may be 16 for 1000 fractalons, for green there may be 380 for a cube of 8000 fractalons) the cosmological constant becomes n for xyz, they will be grouped together in a bubble of fractalons that will spin like a vortex of Schrodinger equation. Gravity without matter = LIGHT. It can be reflected if the architecture between the reflective cubic organization of the vacuum and the intermediate allows it (DSE). When captured depending on the number of additional fractalons it contains, this package be integrated in others fractalons aroun the atom, will ionize an atom producing electrostatic fractalons therefore electricity potential through the photoelectric effect. This Schrodinger bubble does not advance in the ideal vacuum, it disorganizes and reorganizes and finally we will find the n additional fractalons on the other side of the ideal vacuum through Newton's pendulum mechanism and domino effect (Fractalon part 3) Depending on the amount of additional fractalons it contains, it lacks the energy necessary to modify one or more layers of the atom that will integrate it in the form of additional electrons fractalons into its structure. If it deionizes, it manages to send them back into space and free up space in the atom exactly like we blow a soap bubble into a soap film.The electrons are effectively ejected from the atom in the form of light (more correctly pushed), decreasing the pressure on the internal layers until equilibrium is reached. Otherwise, all atoms would transform into increasingly larger atoms, up to black holes that would spin at speeds close to the speed of light, transferring individual rotation into group rotation.If we manage to capture them and direct them in a controlled manner between atoms, we obtain electric current.

Schrodinger's light balloon, reflected and refracted

Schrodinger's light balloon, integrated between the other fractalons of a proton increasing the individual cosmological constant, that is, being ionized (high pressure on the lower layers). Increasing the spherical volume of the atom, more layers of ideal vacuum will be transformed into electrons and a peripheral layer with a larger radius will exert beats, precessions on the internal layers, like at a concert in a stadium when the number of people increases on the periphery and exerts pressure on those in the center.

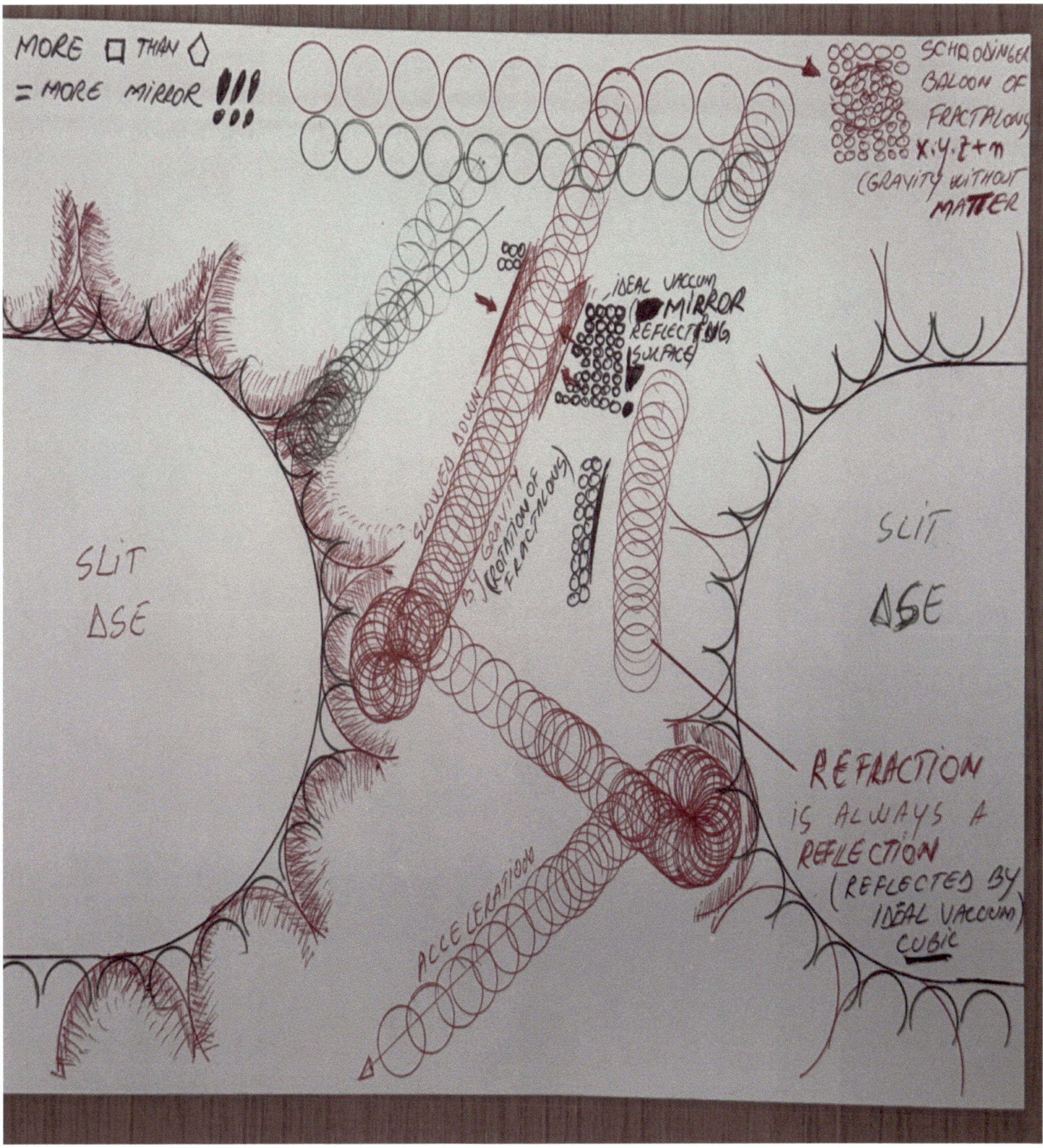

Schrodinger's light balloon, reflected and refracted in DSE

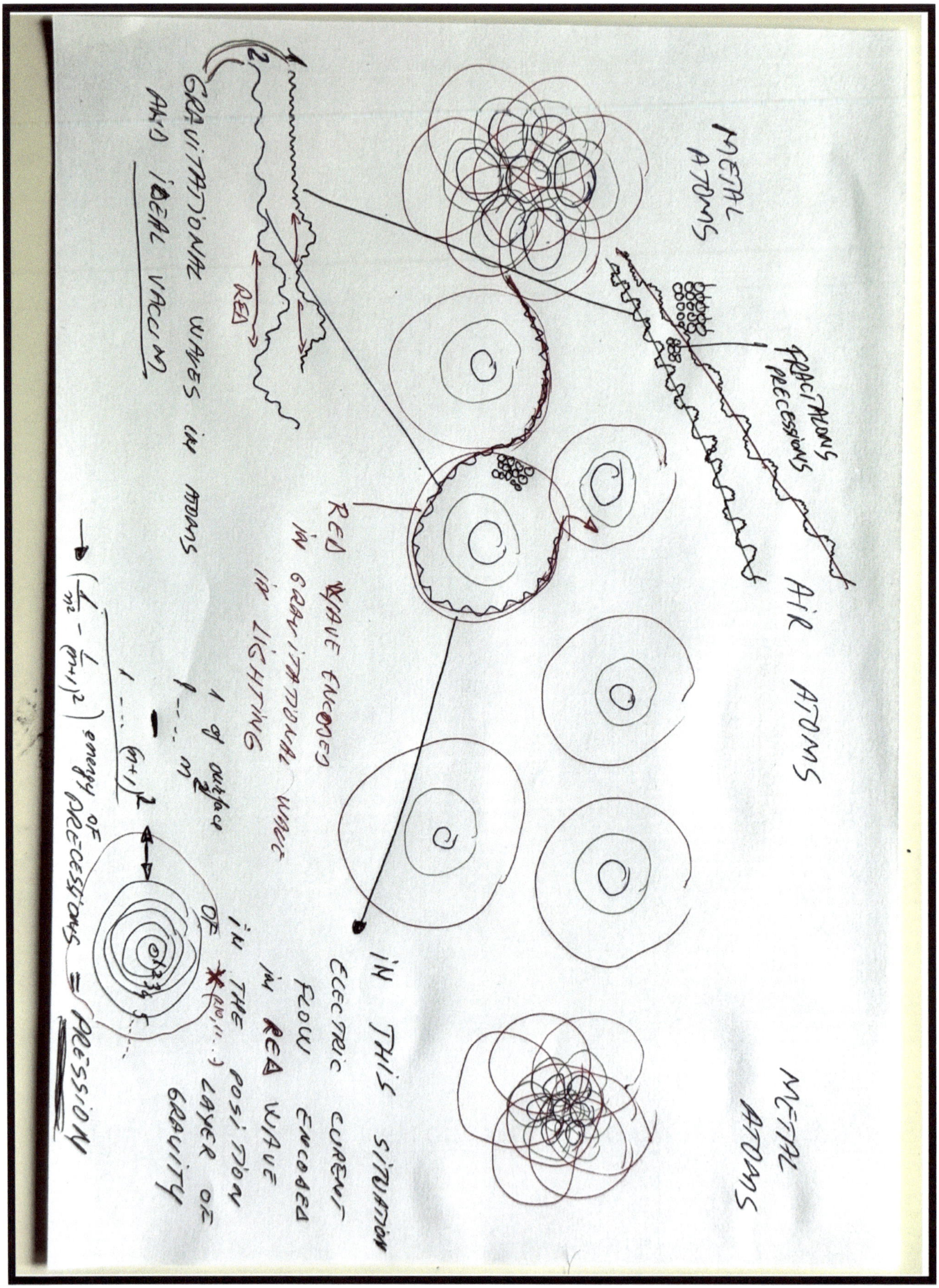
METAL ATOMS
AIR ATOMS
METAL ATOMS
FRACTALONS PRECESSIONS
RED
RED WAVE ENCODED IN GRAVITATIONAL WAVE IN LIGHTING
IN THIS SITUATION ELECTRIC CURENT FLOW ENCODED IN RED WAVE IN THE POSITION OF LAYER OF GRAVITY
2. GRAVITATIONAL WAVES IN ATOMS AND IDEAL VACUUM
energy of PRECESSIONS = PRESSION

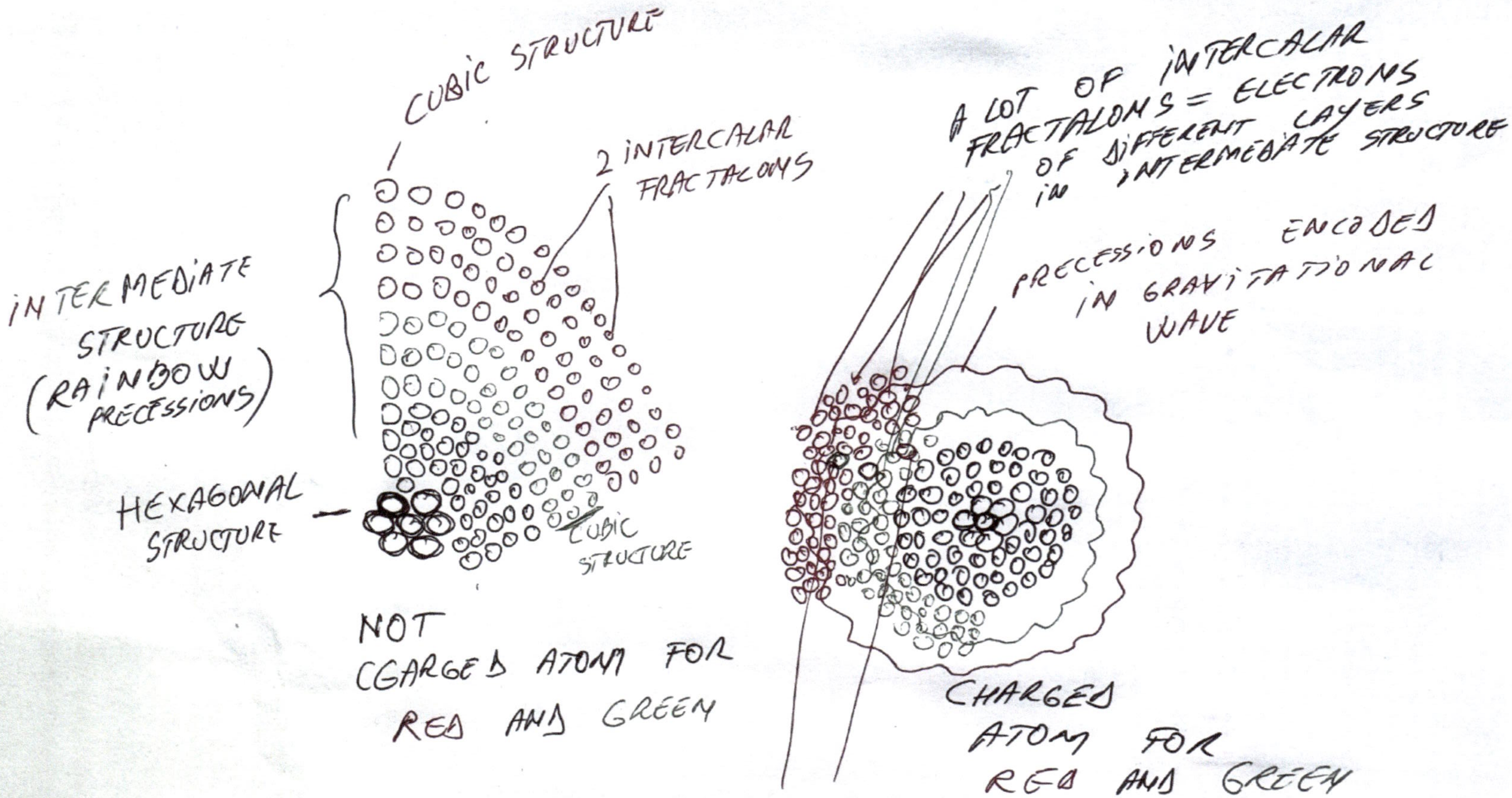

SURACE LAYER

$\pi R^2 \Rightarrow R = 1, 2, 3, 4, 5, 6, 7, (8), 9$

SURFACE LAYER $R = 1, 2, 4, (5), 6, 7, 8$

one of $\pi 8^2$

one of $\pi 5^2$

difference of precessions PRESION $\frac{1}{\pi 5^2} - \frac{1}{\pi 8^2}$

Neutral atoms and ionized atoms, electrostatically charged

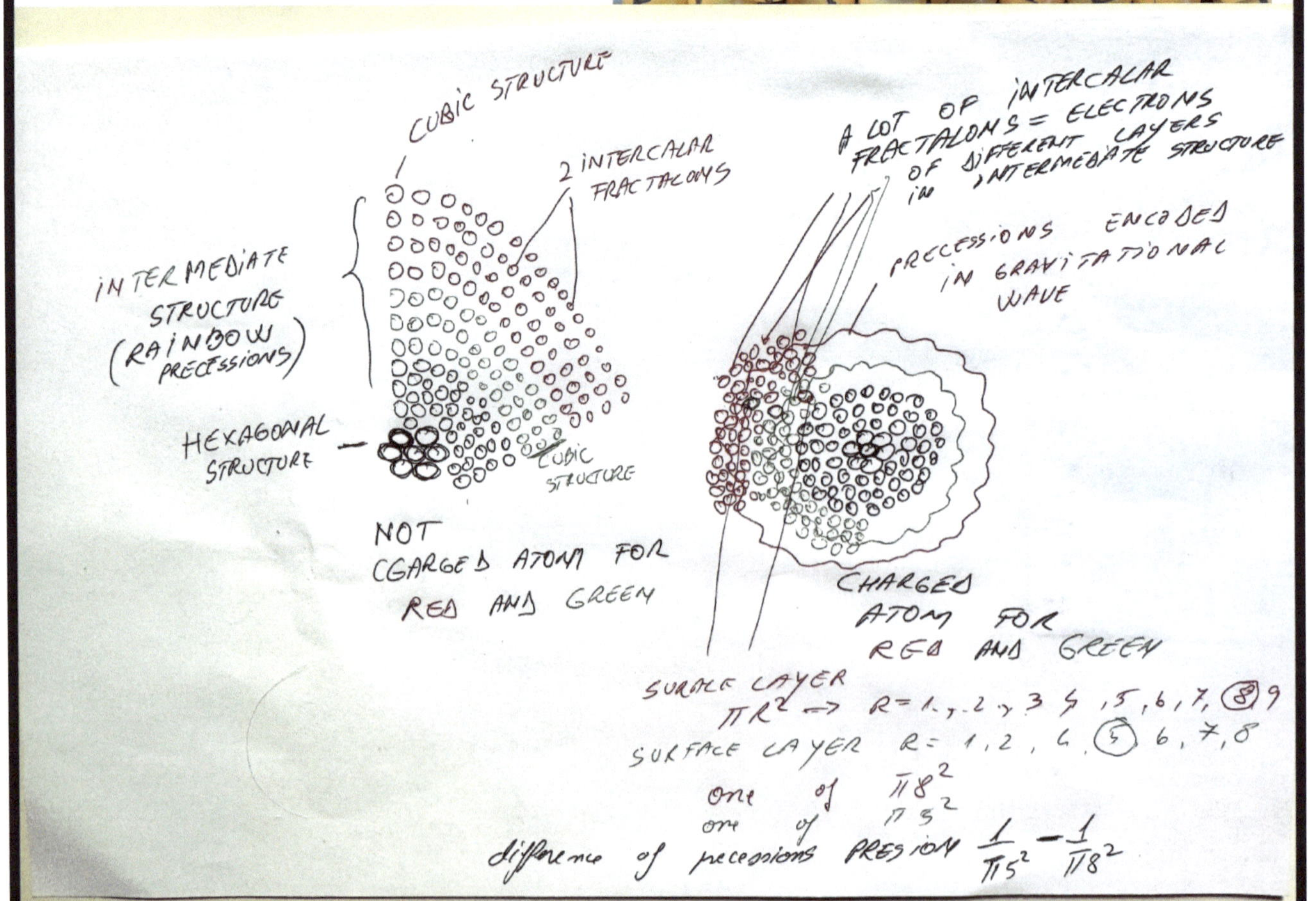

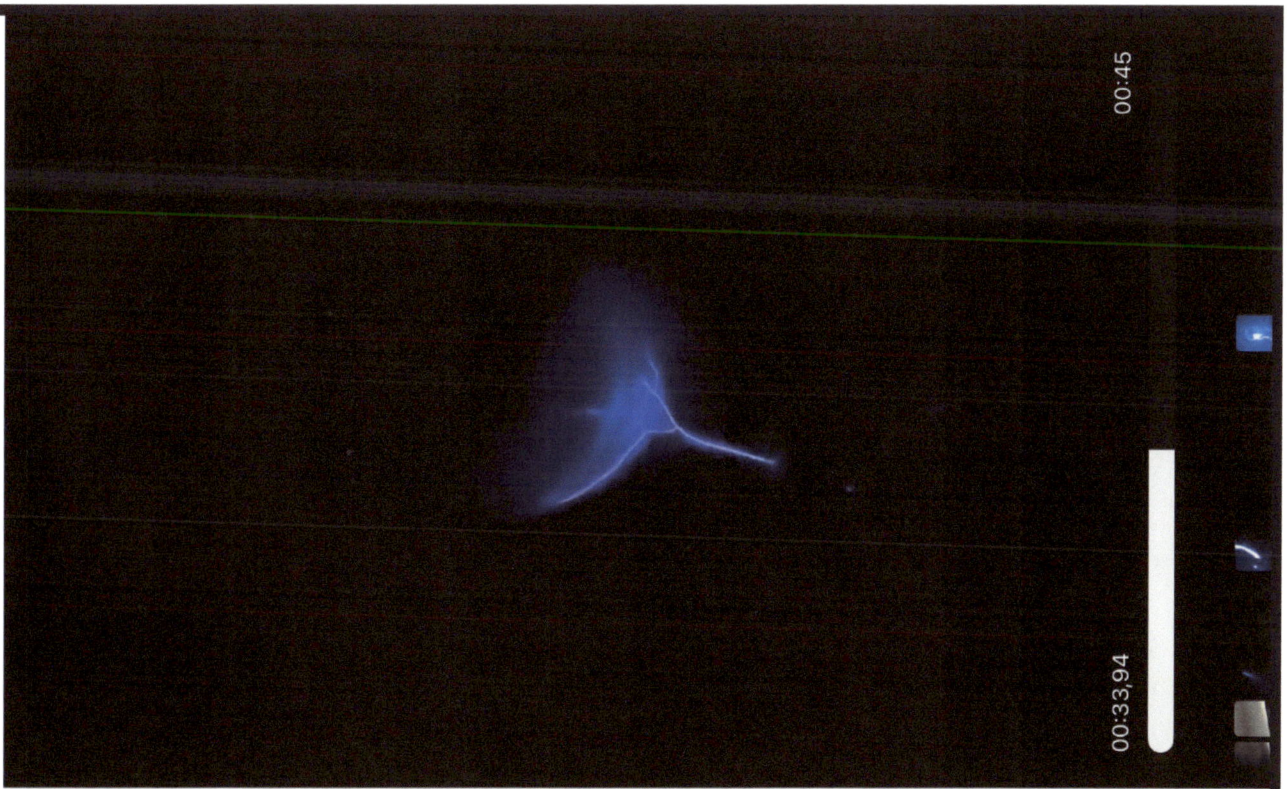

Ampere's law, electric currents in the same direction attract and in different directions repel, for this if we cover the metal bubble with a glass we will obtain rivers of currents that gather in a river, focus, are reflected by the ideal vacuum which will never accept either Schrodinger's light balloons or matter (we are pushed through our electrons by the ideal vacuum towards matter with greater gravity)

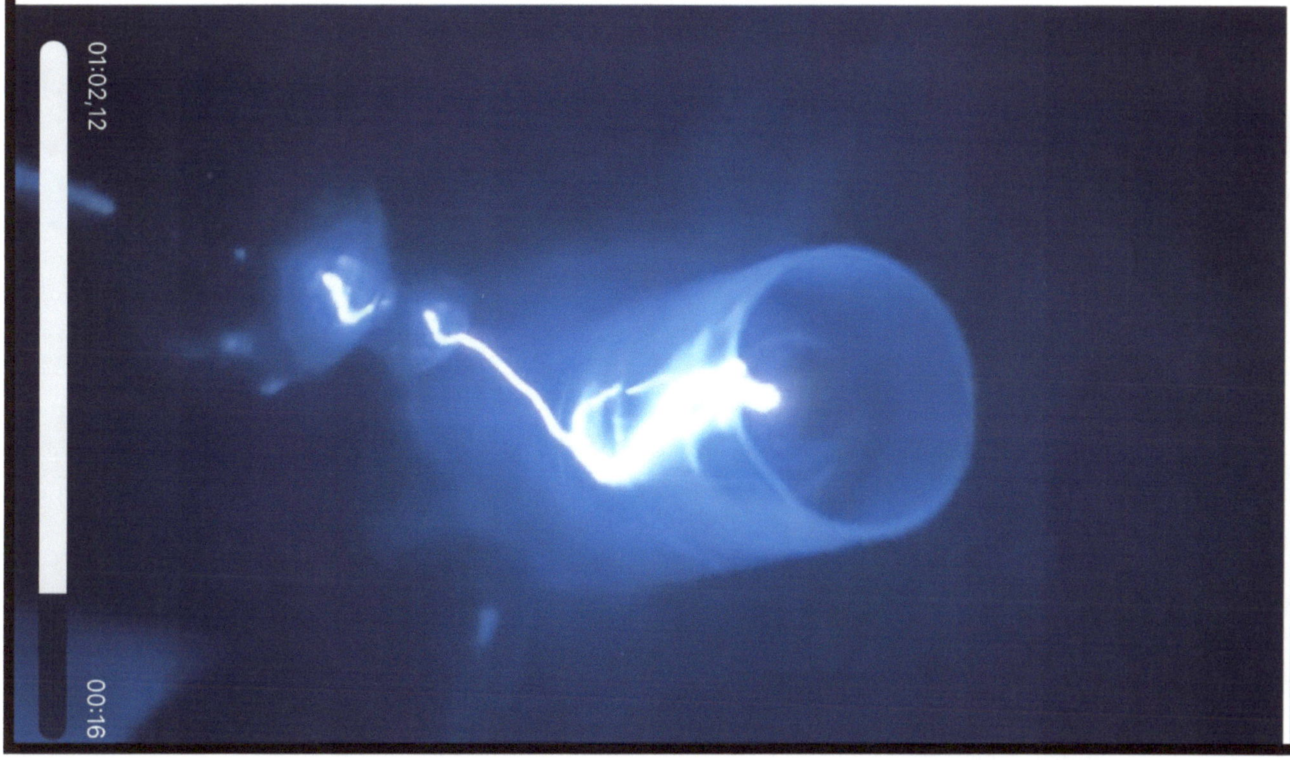

The wavelength of different colors ,like of the sea in the photo doesn't show us the sea is a wave ,the speed of the real wave is the speed of sound and the speed of gravity.Wavelength of light is just encoded frequencies in the wave of gravity.And the sea wave is encoded frequencies in the speed of sound and gravity of the water.But the only two real waves in the photo are gravity and sound waves,and they are not visibles In the photo.
To understand Amper law like in the video of lightingstrike , you apply again my advice to solve the global equations and Fractalon unified theory so the balls pool and you understand that :you must imagine electrons like I say in Fractalon first part book ,like boats ,imagine a lot of boats like a chain boats navigate at 10 meters one of other, first in same ways and second in oposite direction ,what kind of water wave (the ether)they will cross,? and you understand why current in same direction attract the wires and why currents in oposite direction repeal the wires

 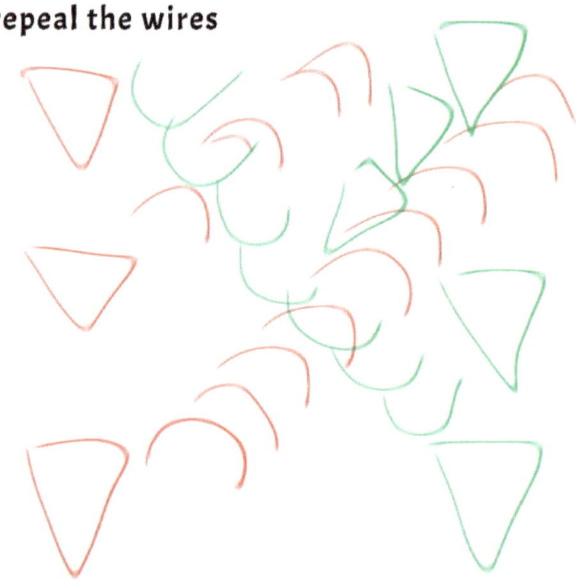

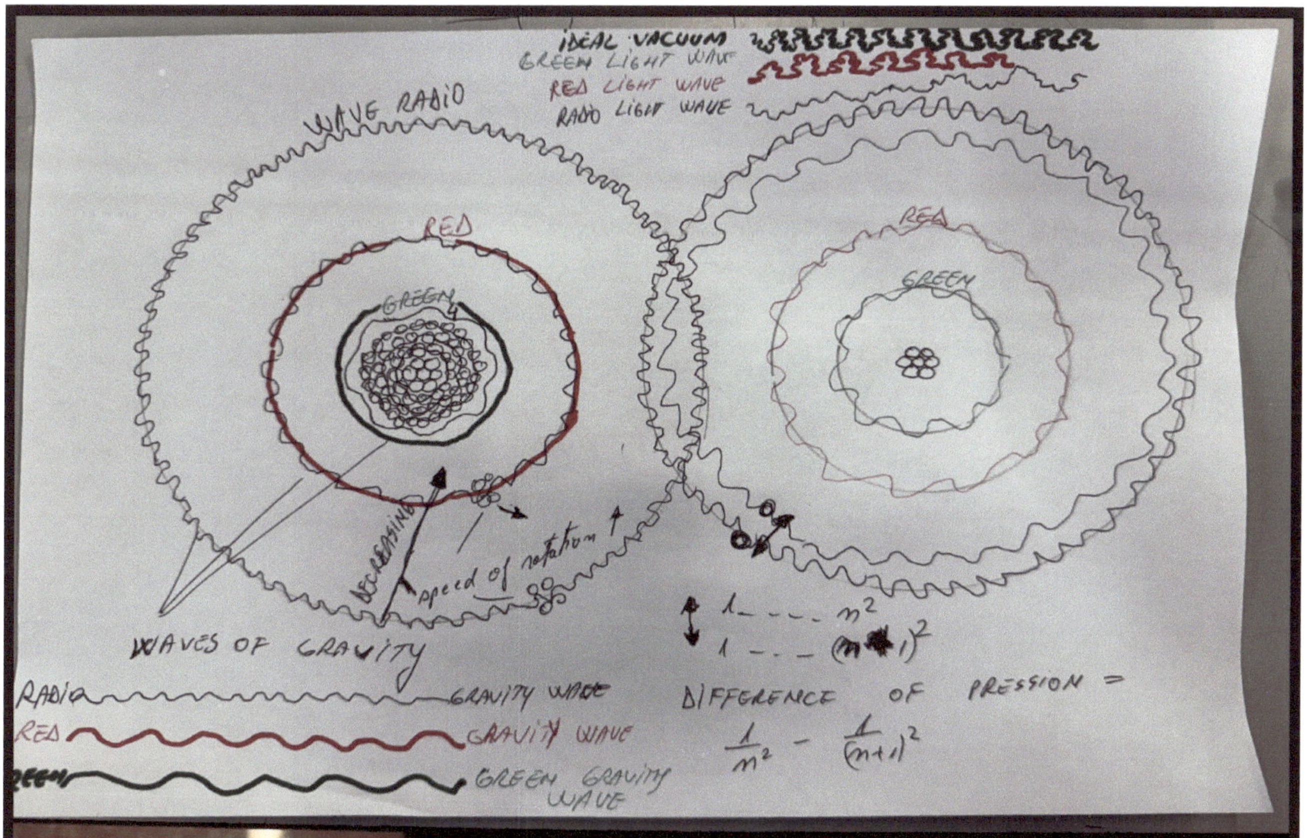

When a series of precessions starts from the green layer for example, it will have a certain frequency compared to the rotation of fractalons in the ideal vacuum, quite low, but as it passes from one layer to the other, this precessional disturbance in Schrodinger's light bubble for green light, will be taken over by fractalons with a higher rotation speed, so it will be accelerated in speed and automatically and in frequency to achieve its double 8, exactly like a series of Fourier transforms, it will store more energy, equivalent to the sum of the energy differences between the layers traversed.

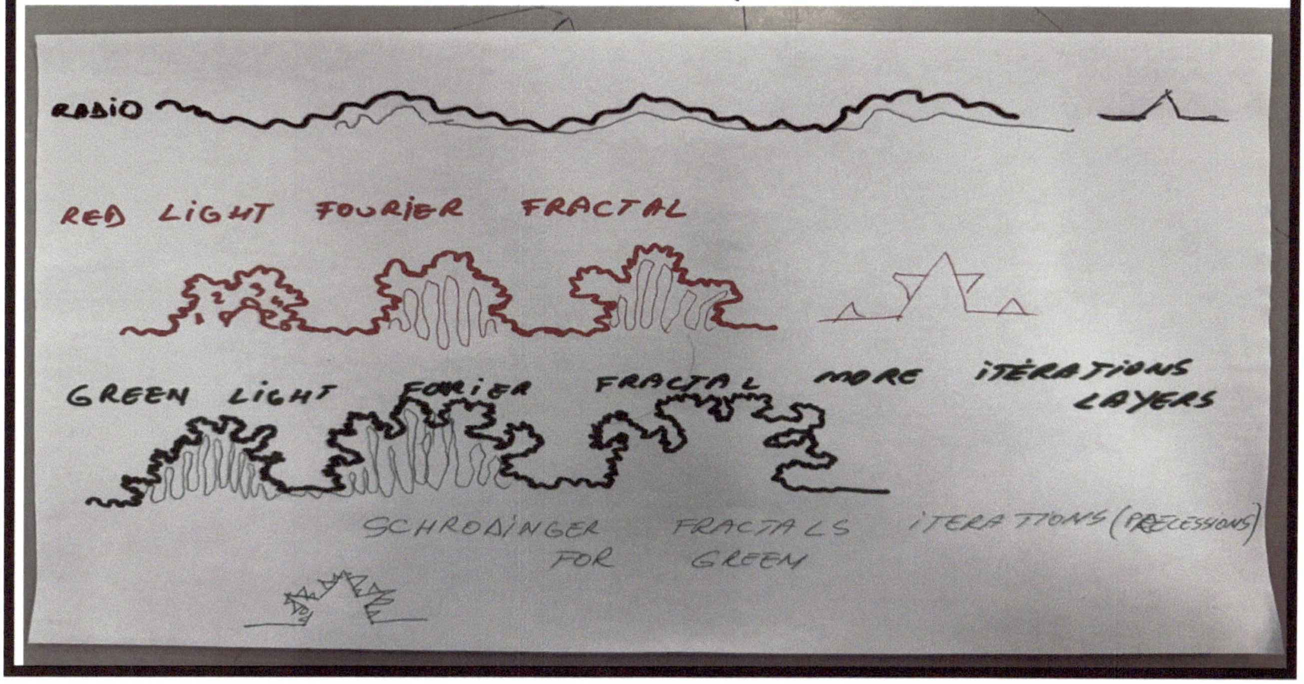

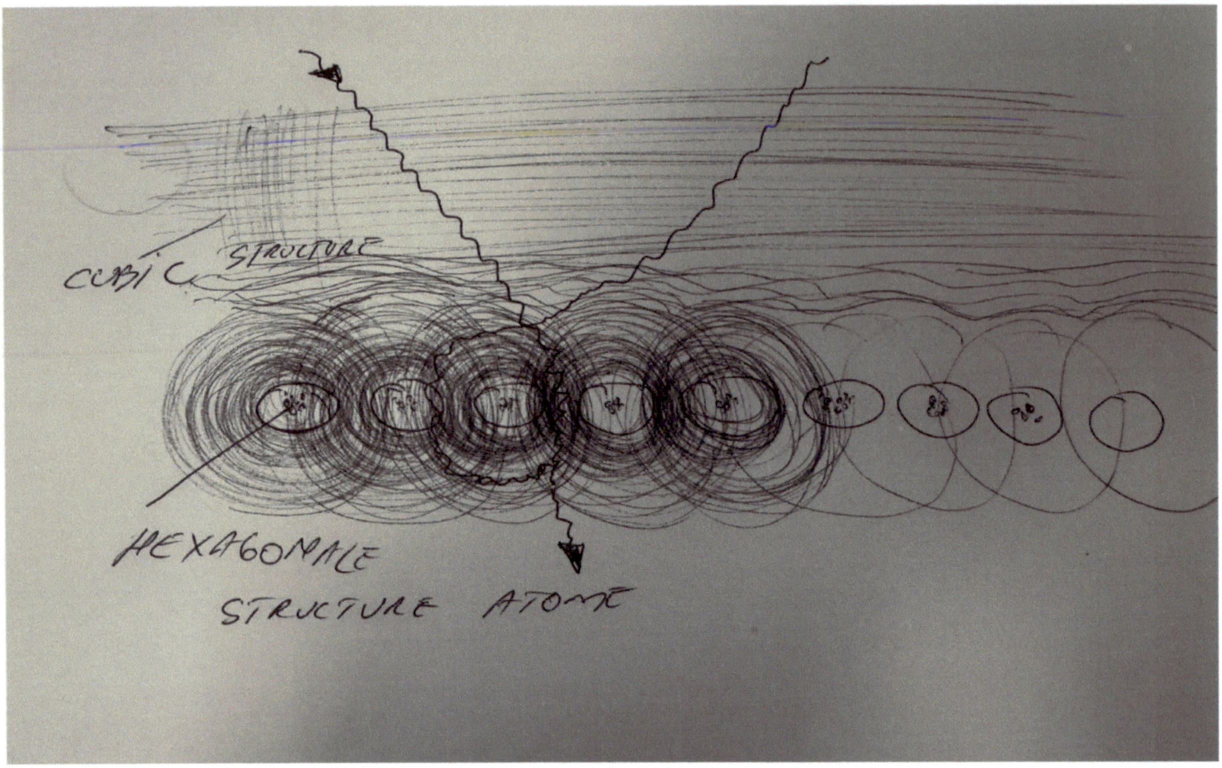

Reflection and "refraction" are similar to the passage of comets in the solar system, only the gravity that will affect them is without matter, it is in Laplace points, light is light enough as gravity to be turned by intercalary fractalons from Laplace points, that is, it will turn or change direction before reaching the earth, it is as if an asteroid were turned by the Laplace point 2, each wavelength has a different Laplace point

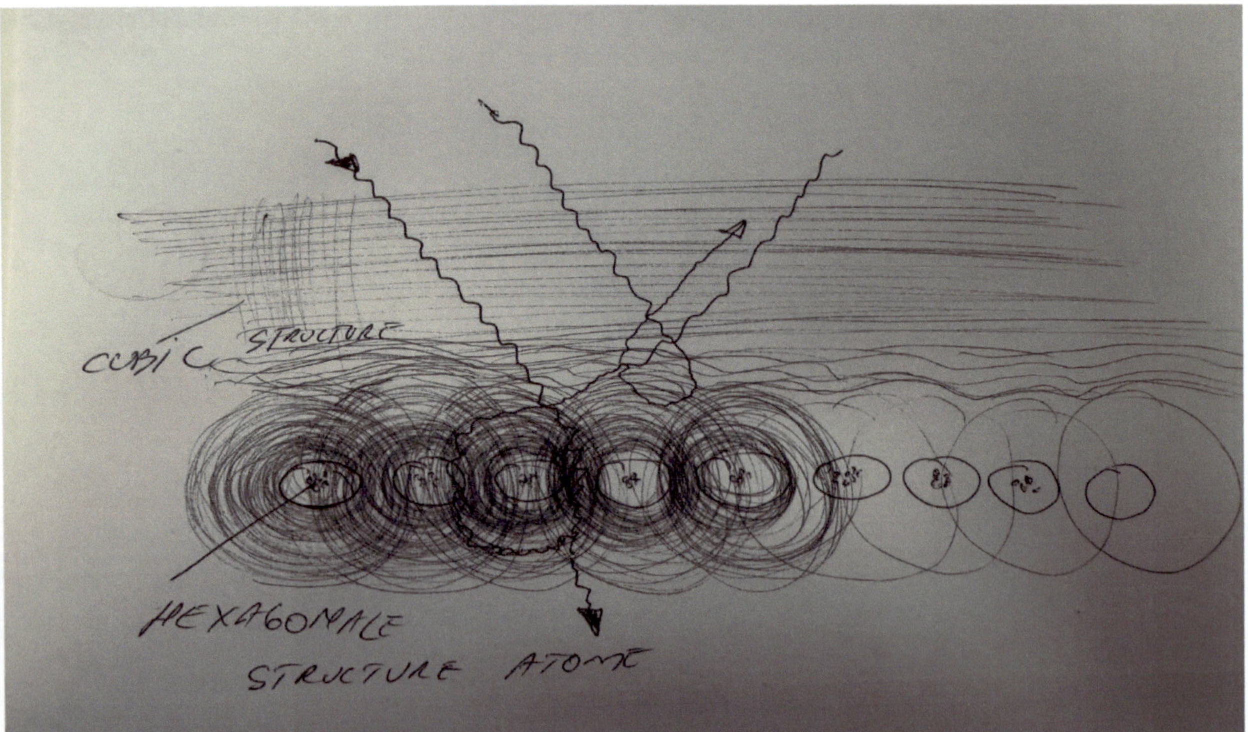

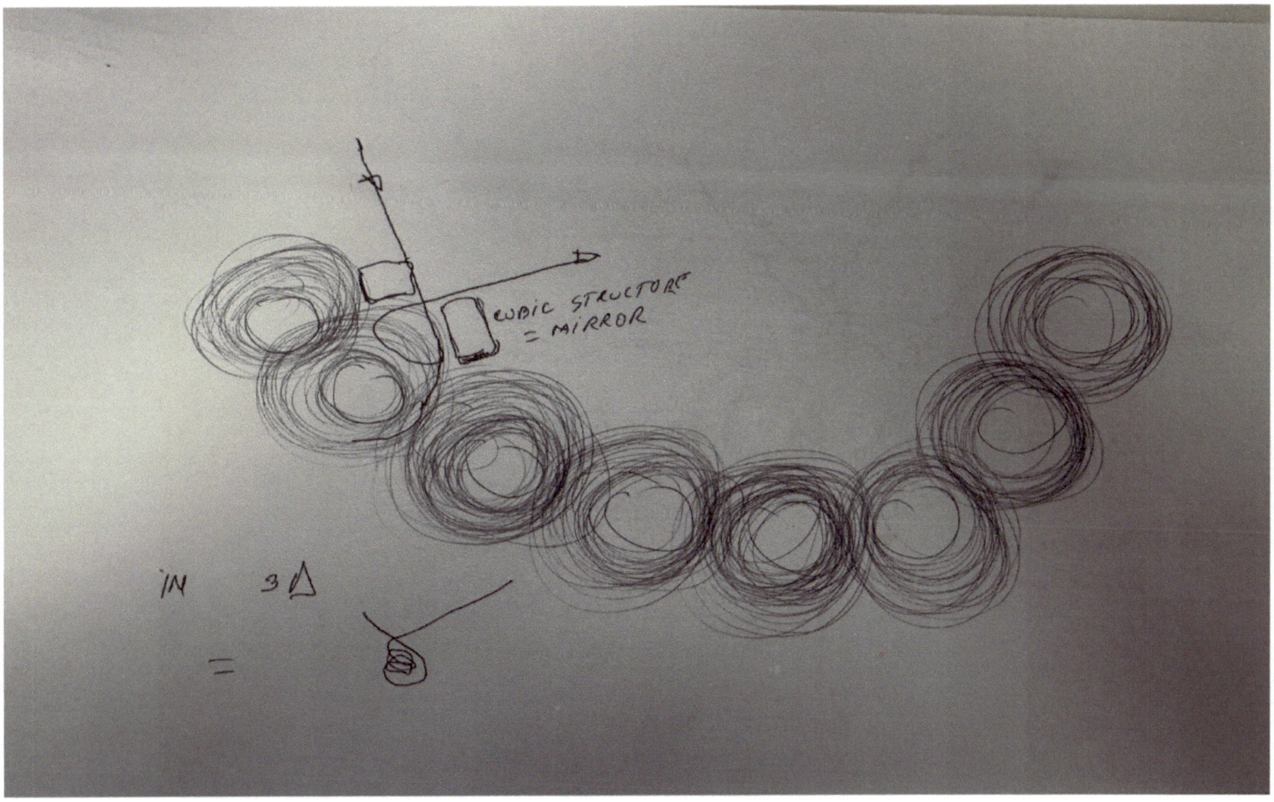

Seen from the outside in 3D it doesn't look like a reflection it looks more like a plane landing in giant circles on a vertical cylindrical runway, with a ball thrown into a conical cup coming back out after a rotation

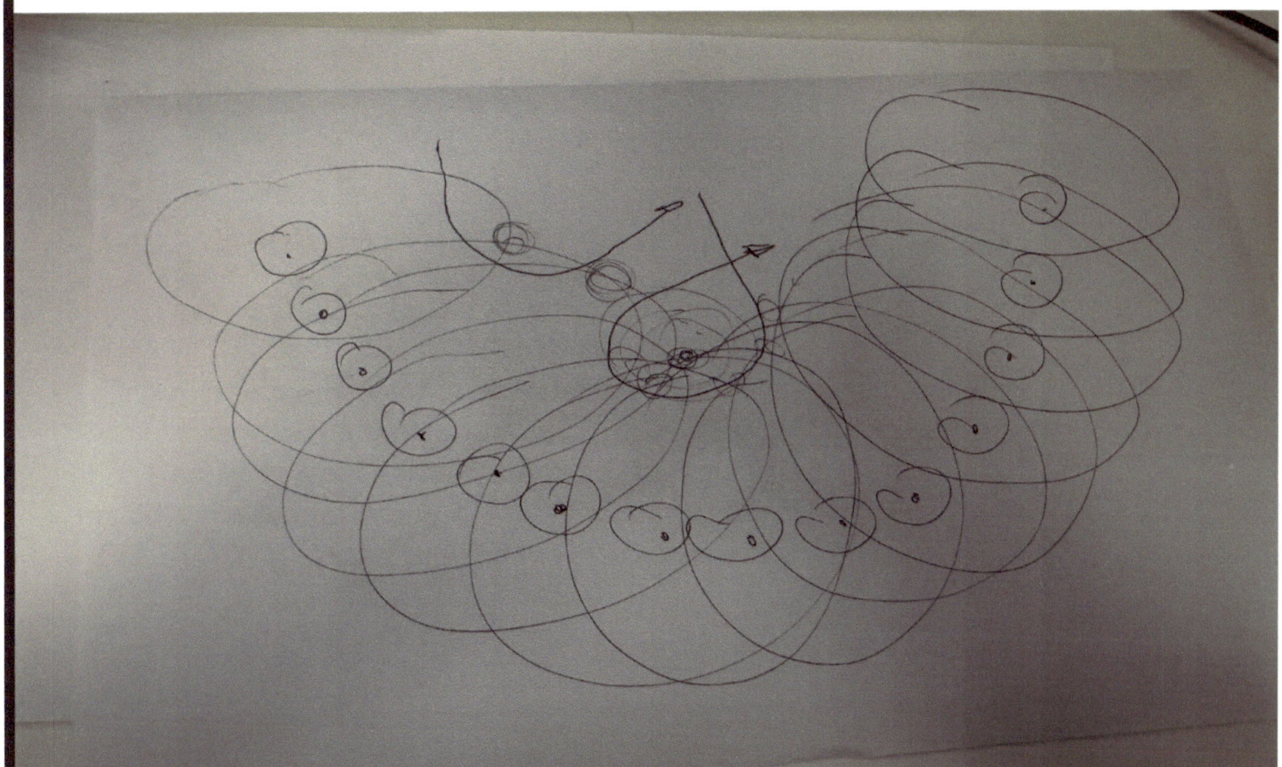

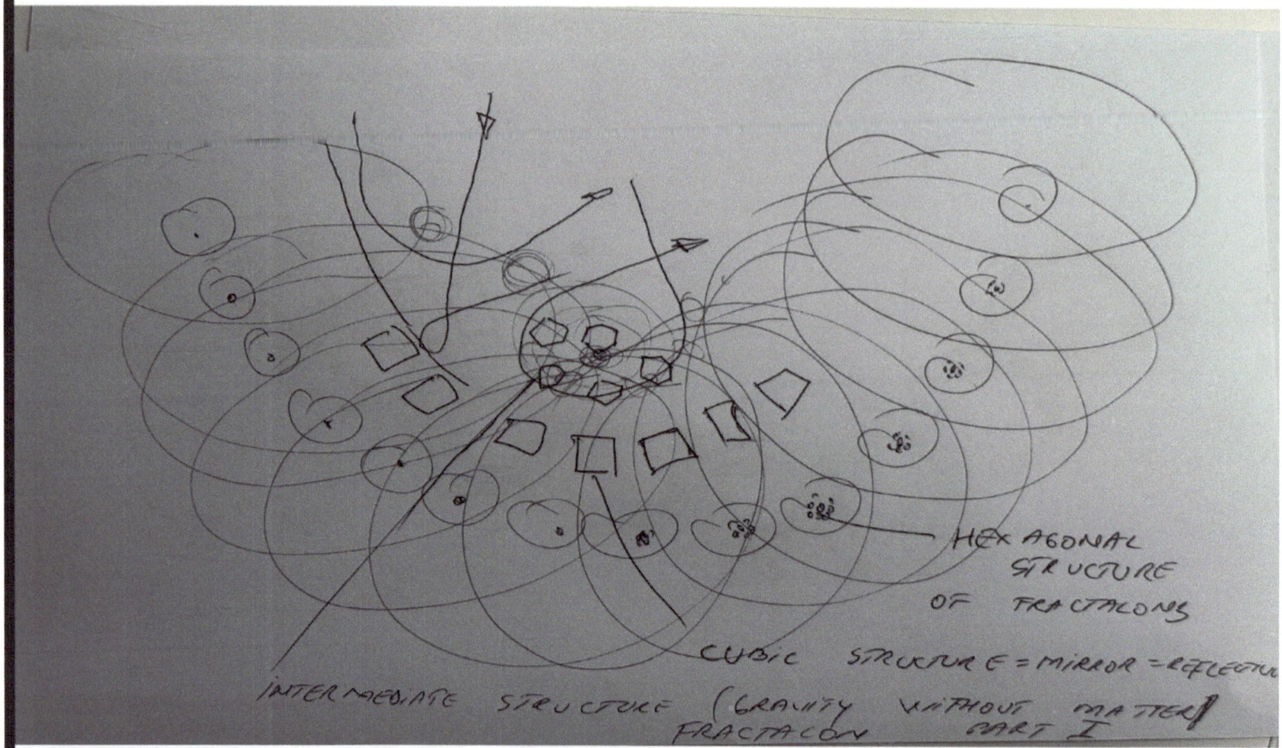

So the light reflects on the vacuum layers around GRAVITY WITHOUT MATTER(FRACTALON part 1), electrons at a distance without very high electrostatic potential, are directly part of the vacuum, of the ether, but are created by matter, like all the other LAPLACE points.But with sufficient gravity to turn the light beam- INTERCALARY FRACTALONS (leptons?)

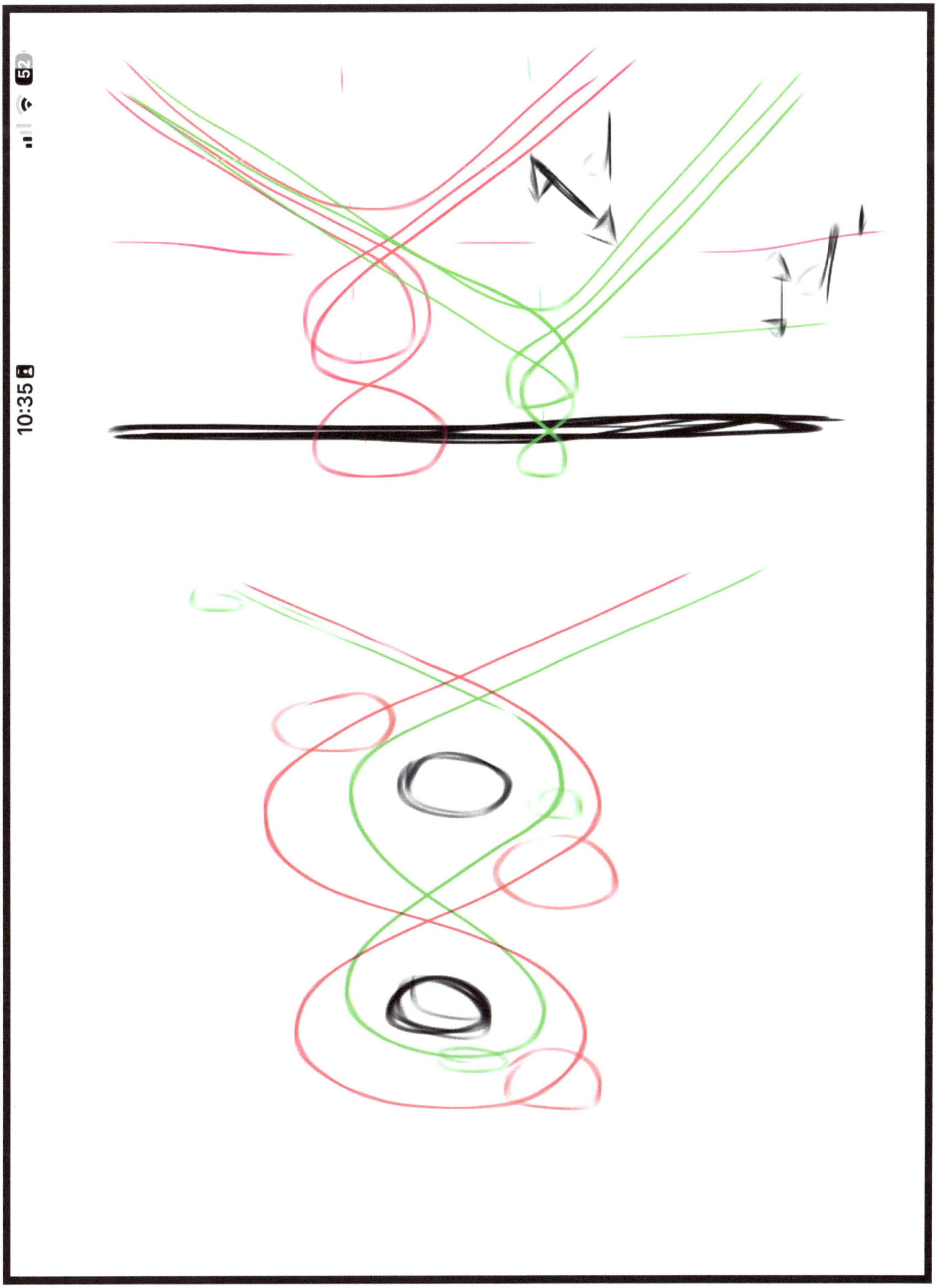
10:35
52

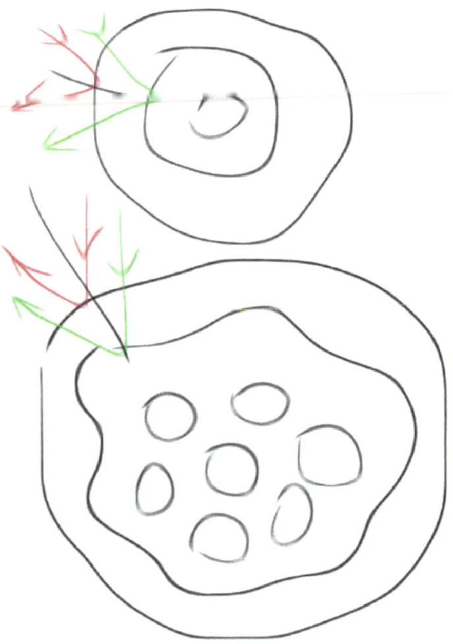

KEY of the DSE experiment

On the same slit in the DSE, the wavelength lights will be reflected at different distances, so the two mirrors are considered to have different diameters for green, red light, etc.

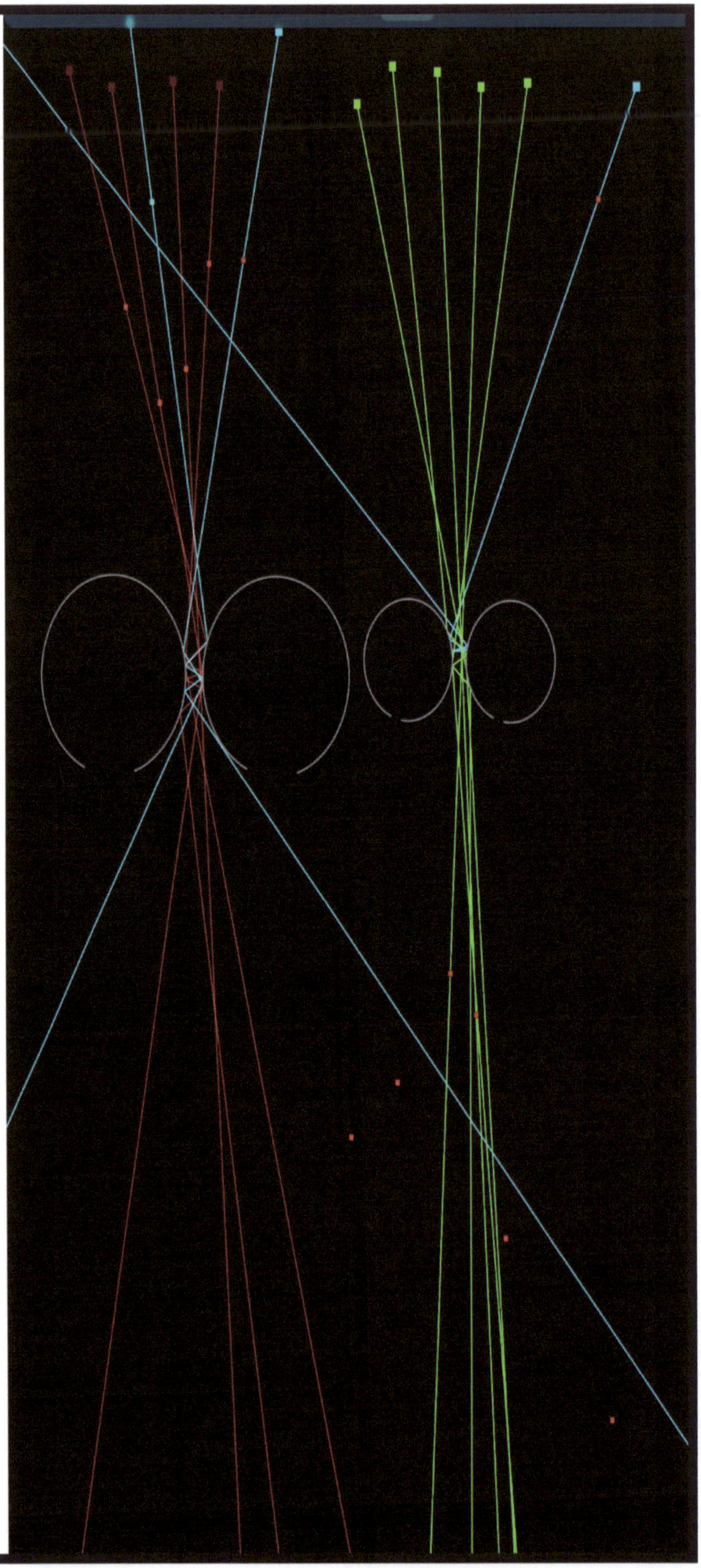

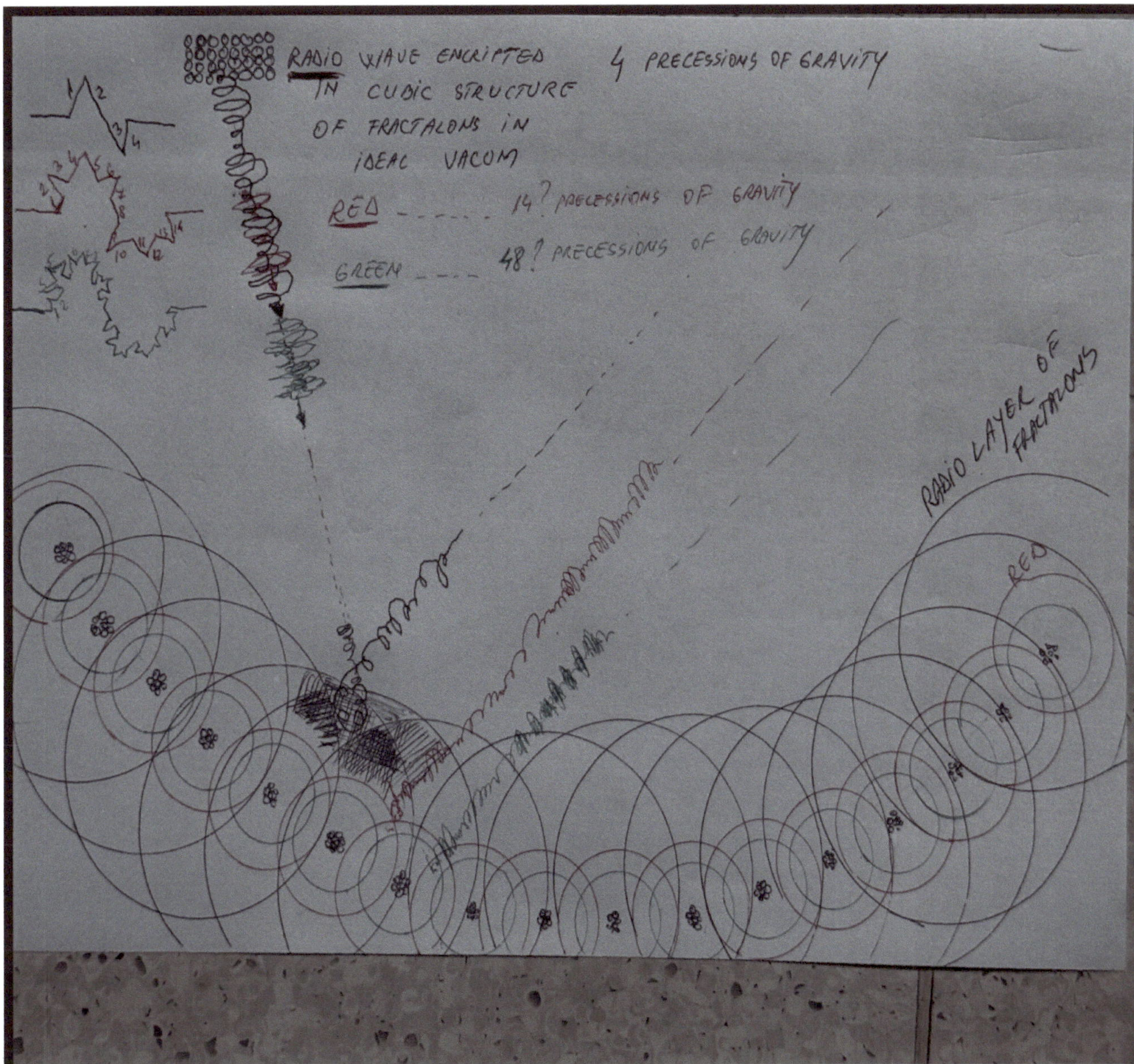

The phase shift that occurs between different wavelengths due to reflection at different distances(on cubic structure of ideal void), in the ether arround Gravity Without Matter from Laplace's Points

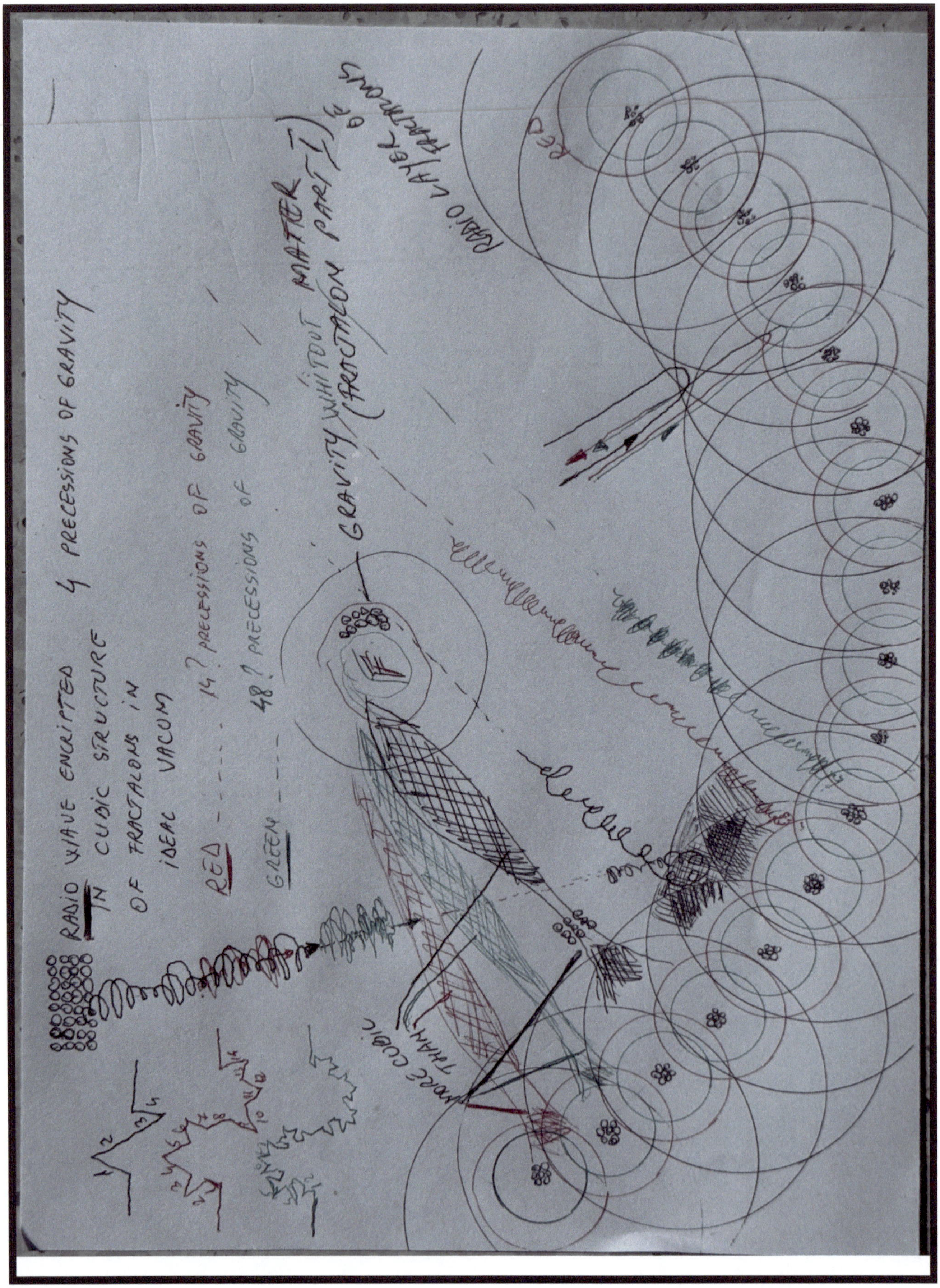
RADIO WAVE ENCRIPTED
IN CUBIC STRUCTURE
OF FRACTALONS IN
IDEAL VACOM
4 PRECESSIONS OF GRAVITY
RED ---- 14? PRECESSIONS OF GRAVITY
GREEN ---- 48? PRECESSIONS OF GRAVITY
GRAVITY WHITOUT MATTER
(FRACTALON PART I)
RADIO LAYER OF FRACTALONS
RED
MORE CUBIC
THAN

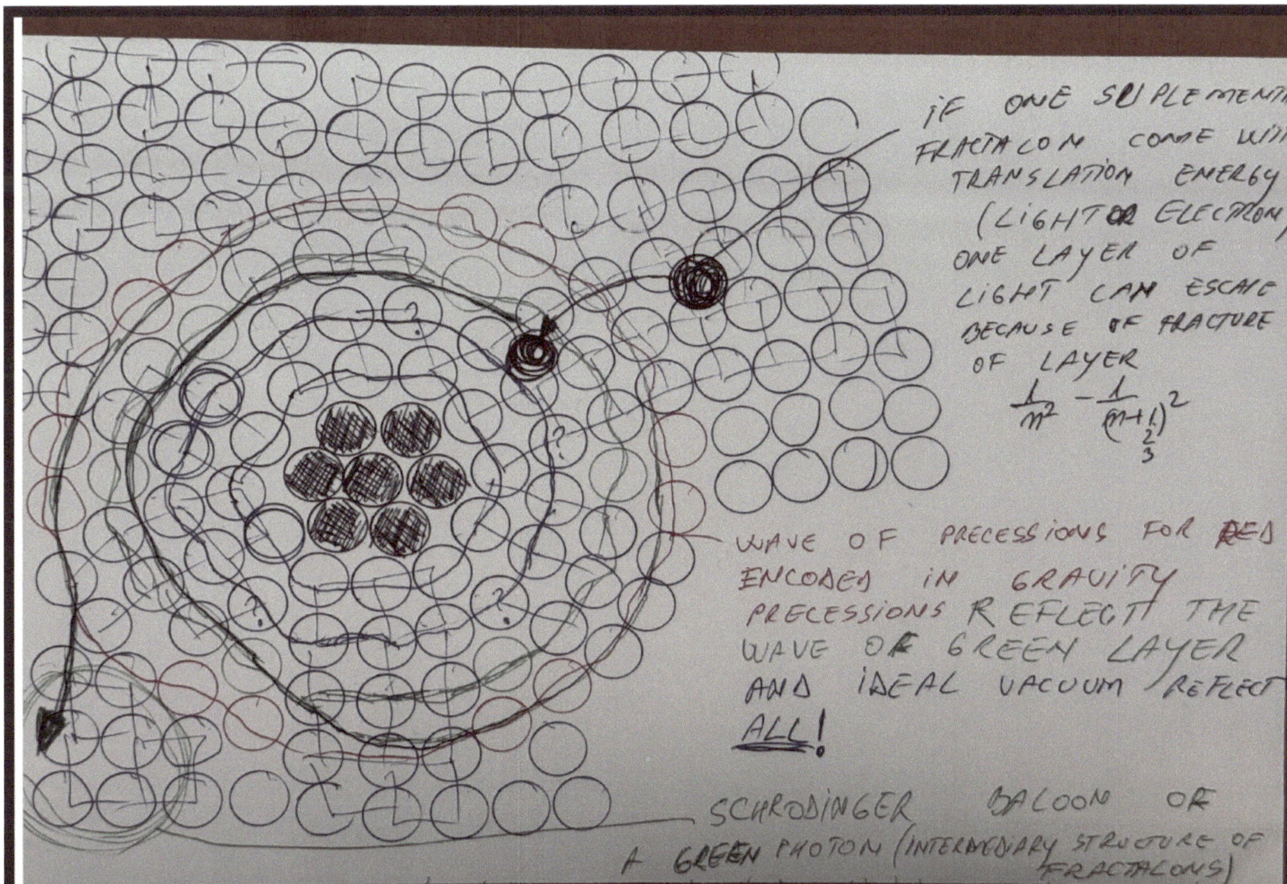

How the translational energy of an electron (when we hit an anvil with a hammer, e.g., electricity or chemical reaction) is transformed into light

This means that the rotation speed will decrease towards it, so at the moment when a second additional Higgs boson is added to the system, the 4 neighbors in the direction of the first will rotate less quickly than the 4 opposite neighbors and the same thing for the other.

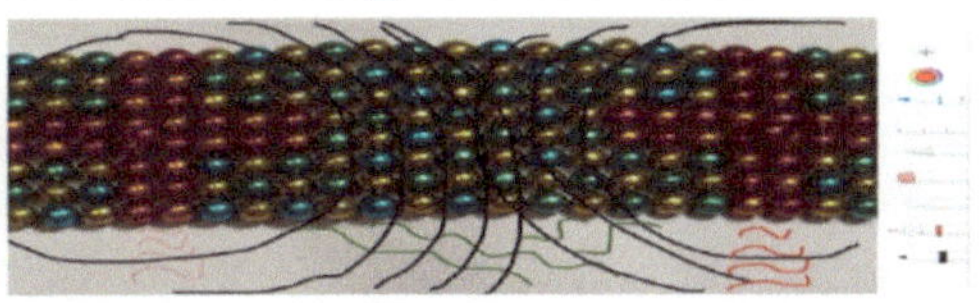

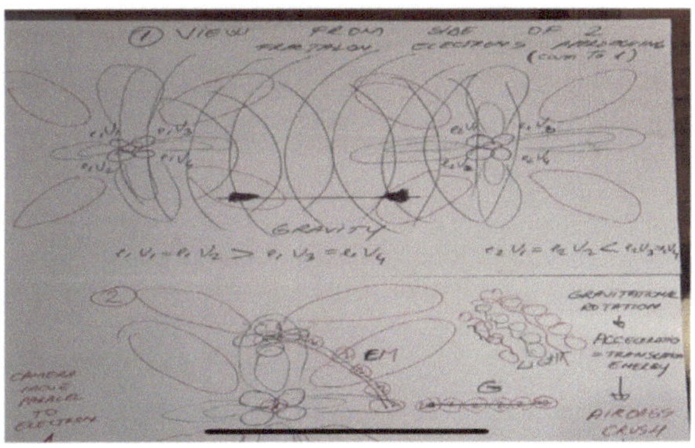

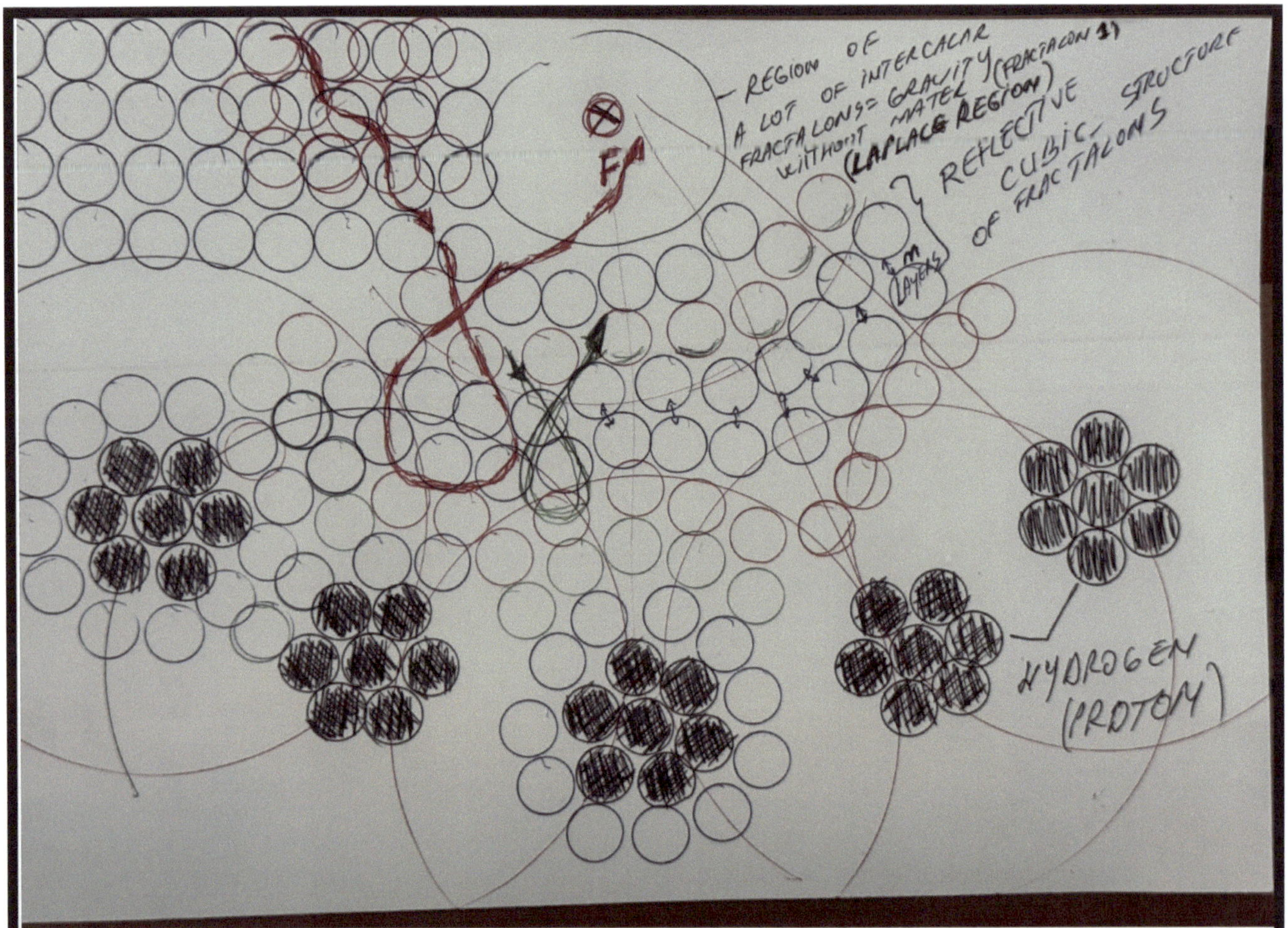

Light creates its own layer of cubic fractalons on which it will be reflected depending on the number of additional fractalons it contains in the quantum (its string of Schrodinger balloons) and the number of intercalated fractalons around its 360 degrees), for this reason on a 0.07 mm wire we can see what happens, we see the mirror image of a surface of 150x650 nm. The real mirror is actually in the ether, formed by fractalons the size of the Higgs boson

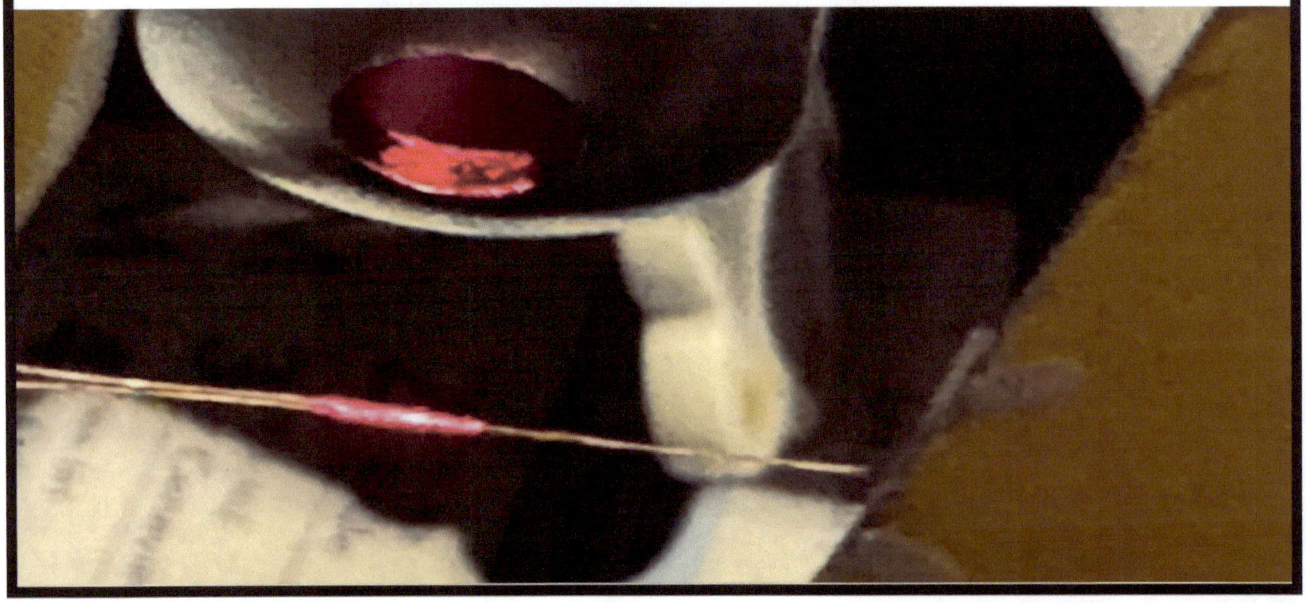

Postulates of modern physics

1.There is at least one real, physical, "palpable" particle and it has a shape and the possibility of contact with its neighbors.
2.Any real particle if it has energy this energy can only be of two types: rotation and translation.
3.A rotation and translation particle has a larger vital space than a rotation particle, which has a vital space greater than or equal to a No Energy particle.

4.A single particle in rotation and translation has a total vital space spherical,regardless of its shape.If this particle is part of a group of particles, the living space may be smaller and shared in time with the others.

5.As the vital space of a group of particles with initial non-zero energy (translation and rotation) diminishes, there is a unique number of particles per unit volume, a unique density at which the particles will self-arrange, in a structure with half the columns spinning in one direction and half in the other, in this structure the particles will have only rotational energy and maximum rotational speed. We can call this structure the IDEAL VACUUM.

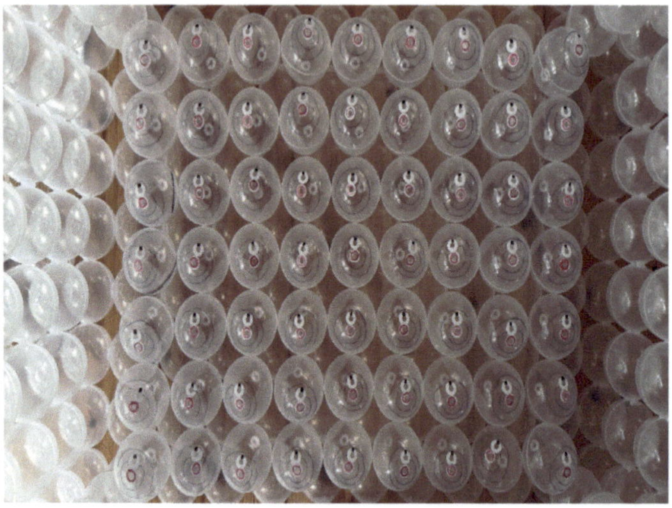

any new particles introduced into the system after reaching cubic organization will generate the appearance of precessions in a group of particles and will contribute to the UNIVERSAL COSMOLOGICAL CONSTANT

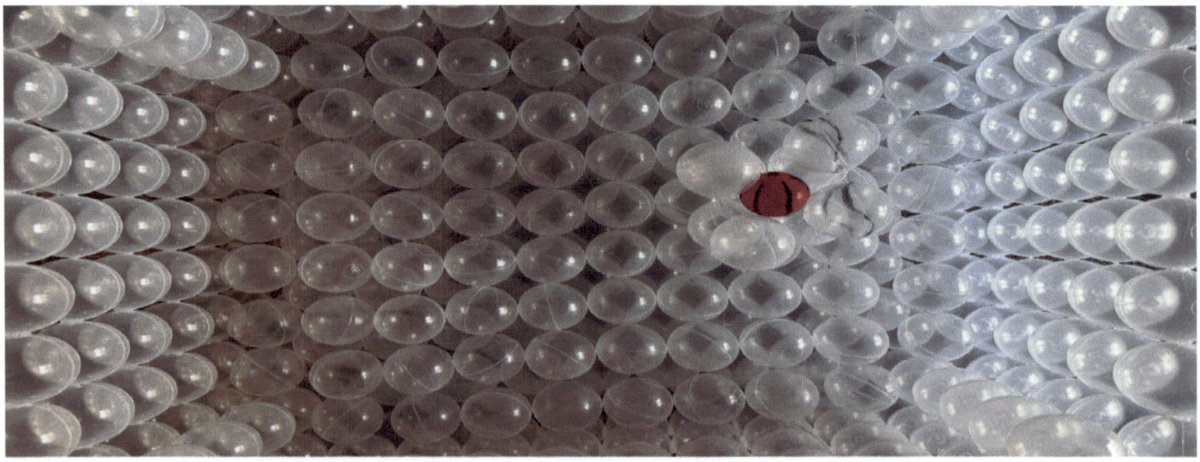

6.Gravity and electromagnetism are not independent phenomena, they are consequences of particles precession. TO OBTAIN GRAVITY AND ELECTROMAGNETISM it is necessary to start from a rotating particle system without gravity and electromagnetism. The strong and weak forces are only pressures exerted on one or a group of particles with a similar translational motion.(fractalon.fr videos)

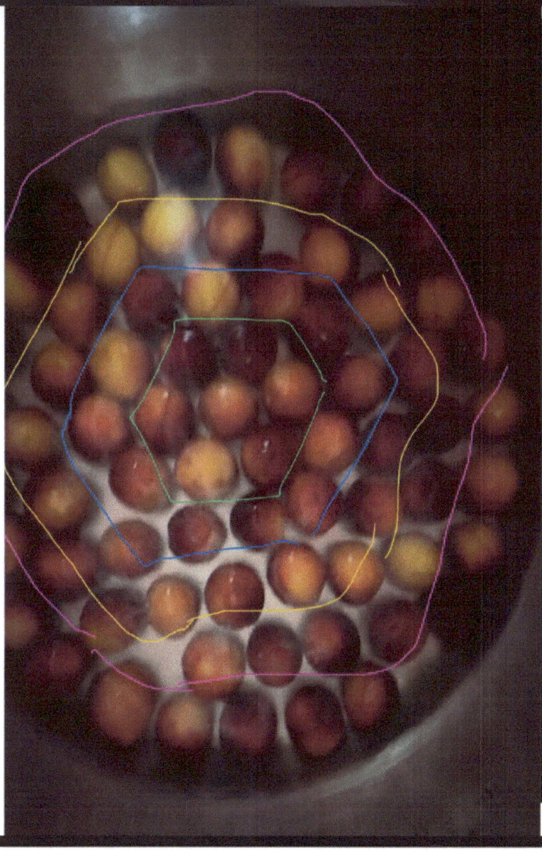

7.In the absence of TRANSLATION ENERGY or CONTACT, there is no transfer of information between particles, so there is no TIME.

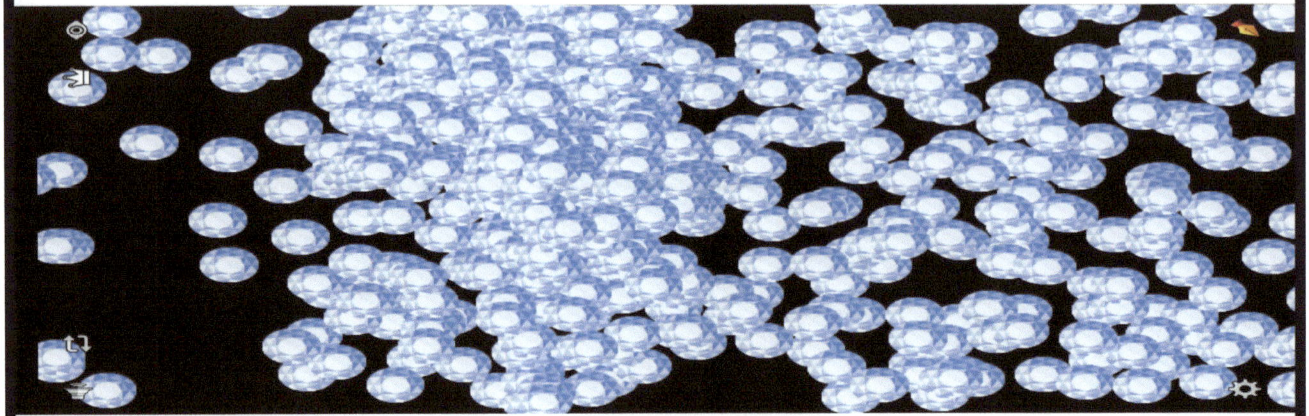

8.All 4 fundamental forces do not exist, they are only consequences of the lateral pressures created by the FRACTALONS precession from outer spheres with a higher number of beats(precessions due to a higher rotation speed in the external layer), to the inner layer with a lower rotation speed of fractalons, so lower number of precessions

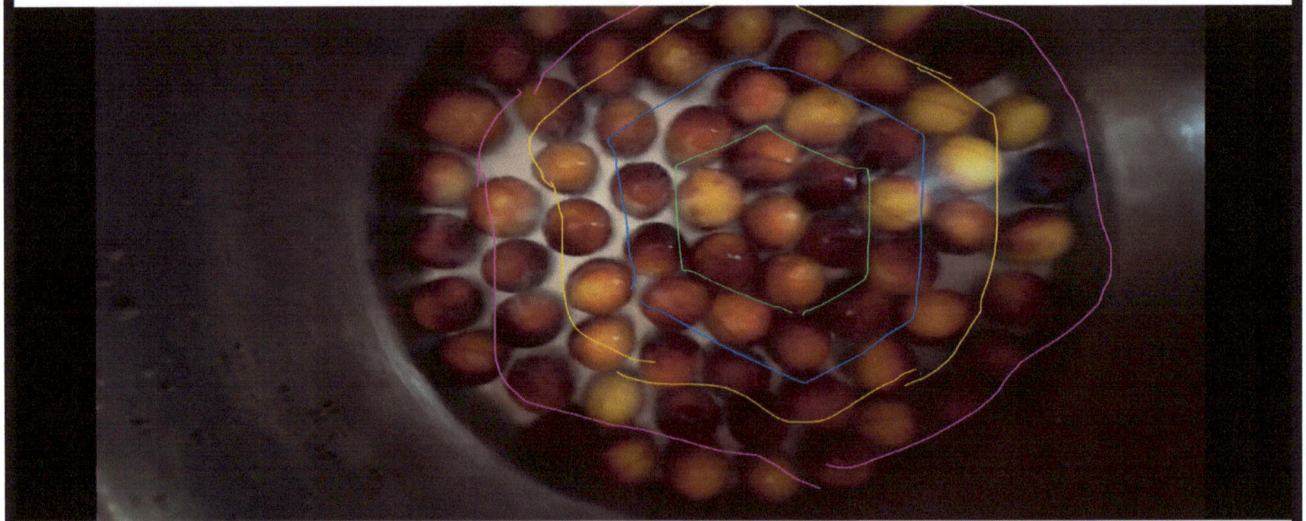

-The difference of pression, of translation energy between two consecutive spherical layers is 1 from layer n with n^2 surface (in fact piXn^2,details,in real univers we don't have spheres or straight lines) to one of (n+1)layer with $(n+1)^2$ surface, so

$$1/n^2-1/(n+1)^2$$

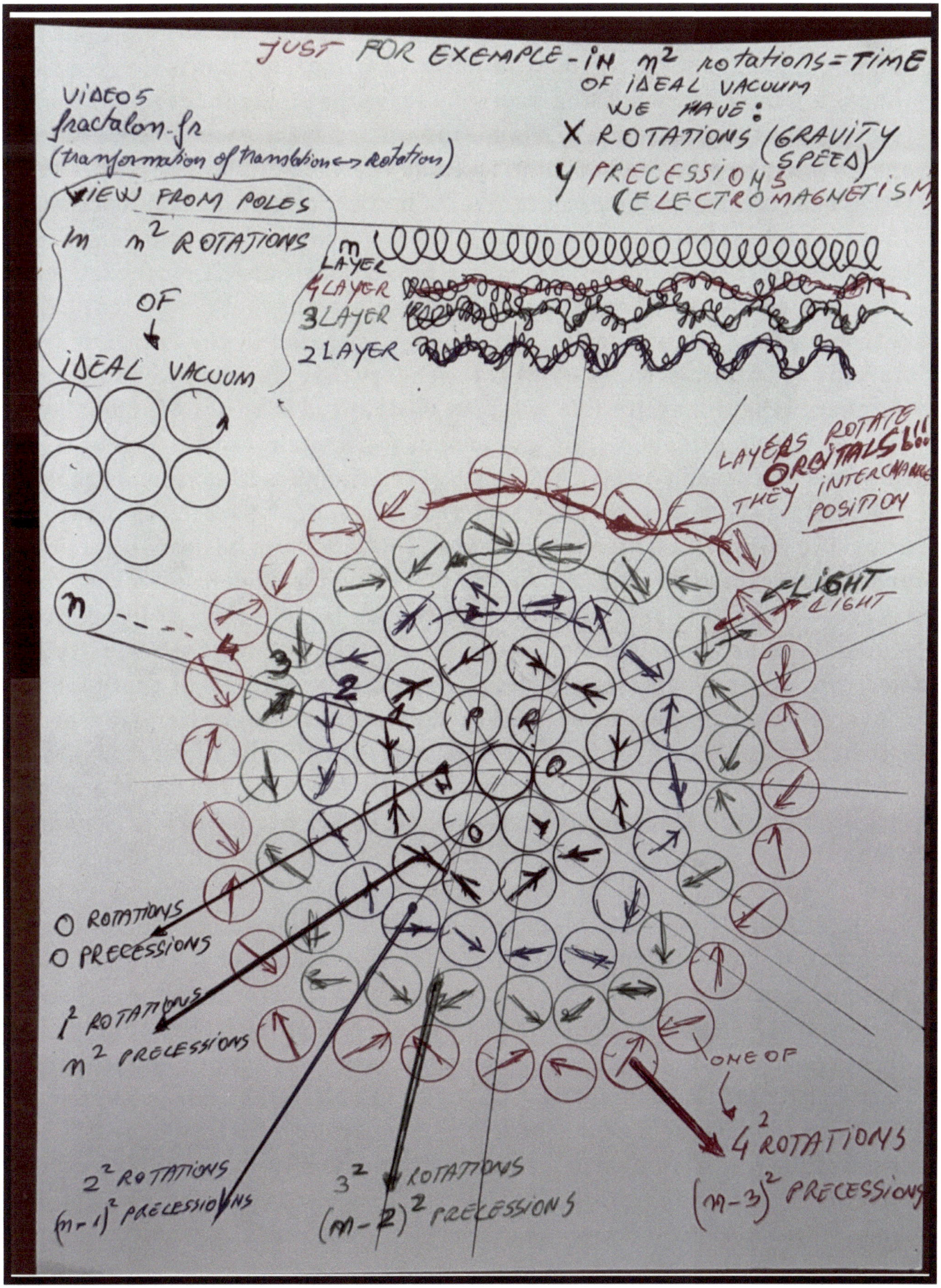
JUST FOR EXEMPLE - IN n^2 rotations = TIME OF IDEAL VACUUM WE HAVE:
X ROTATIONS (GRAVITY SPEED)
Y PRECESSIONS (ELECTROMAGNETISM)
VIDEOS
fractalon.fr
(tranformation of translation ↔ rotation)
VIEW FROM POLES
IN n^2 ROTATIONS
OF
IDEAL VACUUM
n LAYER
4 LAYER
3 LAYER
2 LAYER
LAYERS ROTATE
ORBITALS!!!
THEY INTERCHANGE POSITION
LIGHT
LIGHT
n
4
3
2
1
0 ROTATIONS
0 PRECESSIONS
1^2 ROTATIONS
n^2 PRECESSIONS
2^2 ROTATIONS
$(n-1)^2$ PRECESSIONS
3^2 ROTATIONS
$(n-2)^2$ PRECESSIONS
ONE OF
4^2 ROTATIONS
$(n-3)^2$ PRECESSIONS

All detections of "high energy" bosons are made by accelerating particles to very high speeds and then making impacts. In all measurements they expose the energy values for which the fractalons leave the atoms and resume after a few rotations in groups of 2 or 3 (uud) their place in the vacuum. How much energy must be given to it to put it back in the vacuum. This energy is the latent heat of the vacuum that is recovered in chemical and nuclear reactions. There is a latent heat between VACUUM (Fractalons with rotation and zero translations) ELECTRONS (rotations and translations) and ATOMS (hexagonal structure, only translation) and then for the electrons there is a latent heat between vapor - liquid - solid. This energy is stored in the rotation and translation of the electrons. But it is the energy that must be given to it, not the energy that it has. It's like saying at what speed in an accident you'll go through the windshield. But they're so happy with their research money that this omission doesn't matter. That's why I call them Fractalons in my books, never bosons. Because I described their real existence, not accelerated by this translational energy. Einstein's equation e = mc2 is only valid in one direction, and it's absolutely logical. When you block 7 Fractalons in a proton, you recover 7 times the rotational energy (which is the speed of light) of a Fractalon. When you do fusion between several protons, if the density of electrons around increases you recover all the lost rotational energy. But conversely, we can never recover energy by transforming the iron atom back into helium and then hydrogen - protons and then putting the proton into 7 Fractalons (the uud, 2+2+3).To do this, we must deliver all the latent energy lost in the fusion reactions in stars. W bosons are not particles, they are groups of two particles, two Fractalons, which rotate together, which will disturb space with their translational energy given by humans in the accelerator until they lose this energy and become electrons again. But you are right, if in one day my theory is validated, several Nobel Prizes will be canceled, and I say this at the beginning of my book, and the first one that will fall is that of Morley-Michelson, which no intelligent physicist before Einstein did not recognize, they all implored Einstein with Bohr in front, to recognize that light needs a medium and to move in this direction. In chemical reactions, the same applies: whenever the reaction is exothermic, the rotation speed of the fractalons will decrease, and their density per unit volume will increase because the gyroscopes are less stable. And vice versa, if the reaction is endothermic, the rotation speed of the electrons in the system will increase.

In current physics, they use a single word for everything: POTENTIAL (gravitational, electrostatic, electromagnetic). In my physics, you can visualize this concept: Rotations and translations of fractalons.

The true equation is rather E = -mc2 it means that the energy lost by the vacuum (vacuum called in current physics potential) to synthesize a quantity of matter is equivalent to the part of the vacuum transformed into matter. This energy is lost as gravitational acceleration and as electromagnetic radiation (light, radio, X, gamma etc) in the successive fusion of the vacuum to proton, from the proton to atoms, from atoms to black holes, and in gravitational acceleration. But light slows down or stops completely at the limit of the different layers of gravitation (the refractive index), the only frequency that penetrates to the periphery of protons and neutrons is gamma radiation, so E = mc2 it is rather correct E = mVg2 or Vg is the speed of transmission of gravitation, light moves at this speed, but gravitation having a much higher frequency it is never slowed down or stopped.

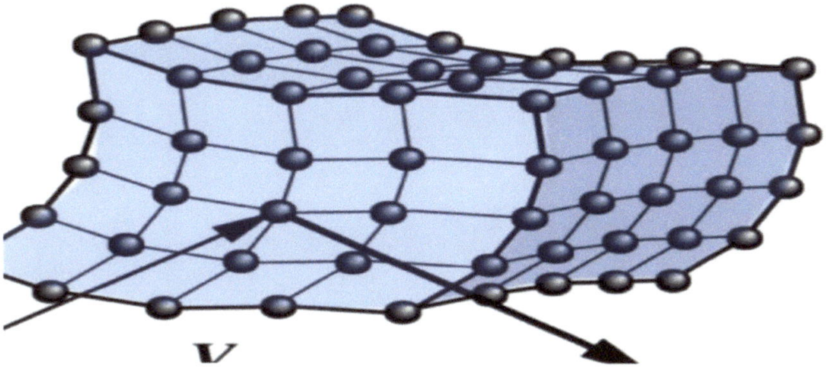

The best conceptualization of ether made with current physics is that of professor Gerard Gremaud the only physicist with whom I managed to have a real exchange of opinions and who deserves at least 2 Nobel prizes for all his work.

The experiment that demonstrates everything I have said in these two pages is that of professor Alex Dmitriyev from the University of St Petersburg who demonstrates that the more energy we deliver to atoms the lighter they become: ENERGY-TEMPERATURE-SOLID-LIQUID-GAS-ETHER

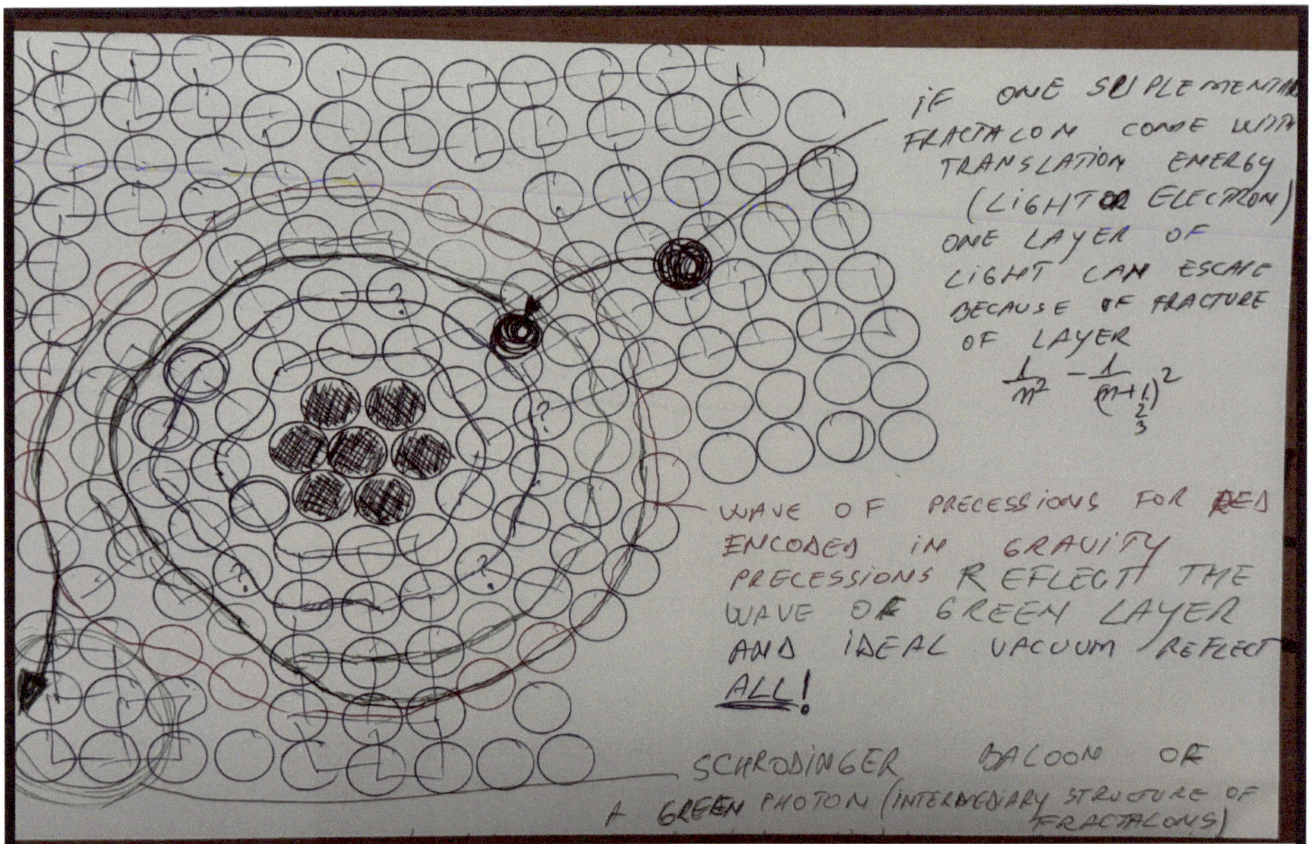
IF ONE SUPLEMENTAR FRACTACON COME WITH TRANSLATION ENERGY (LIGHT OR ELECTRON) ONE LAYER OF LIGHT CAN ESCAPE BECAUSE OF FRACTURE OF LAYER
$\frac{1}{n^2} - \frac{1}{(n+1)^2}$
WAVE OF PRECESSIONS FOR RED ENCODED IN GRAVITY
PRECESSIONS REFLECT THE WAVE OF GREEN LAYER AND IDEAL VACUUM REFLECT ALL!
SCHRODINGER BALOON OF A GREEN PHOTON (INTERMEDIARY STRUCTURE OF FRACTACONS)

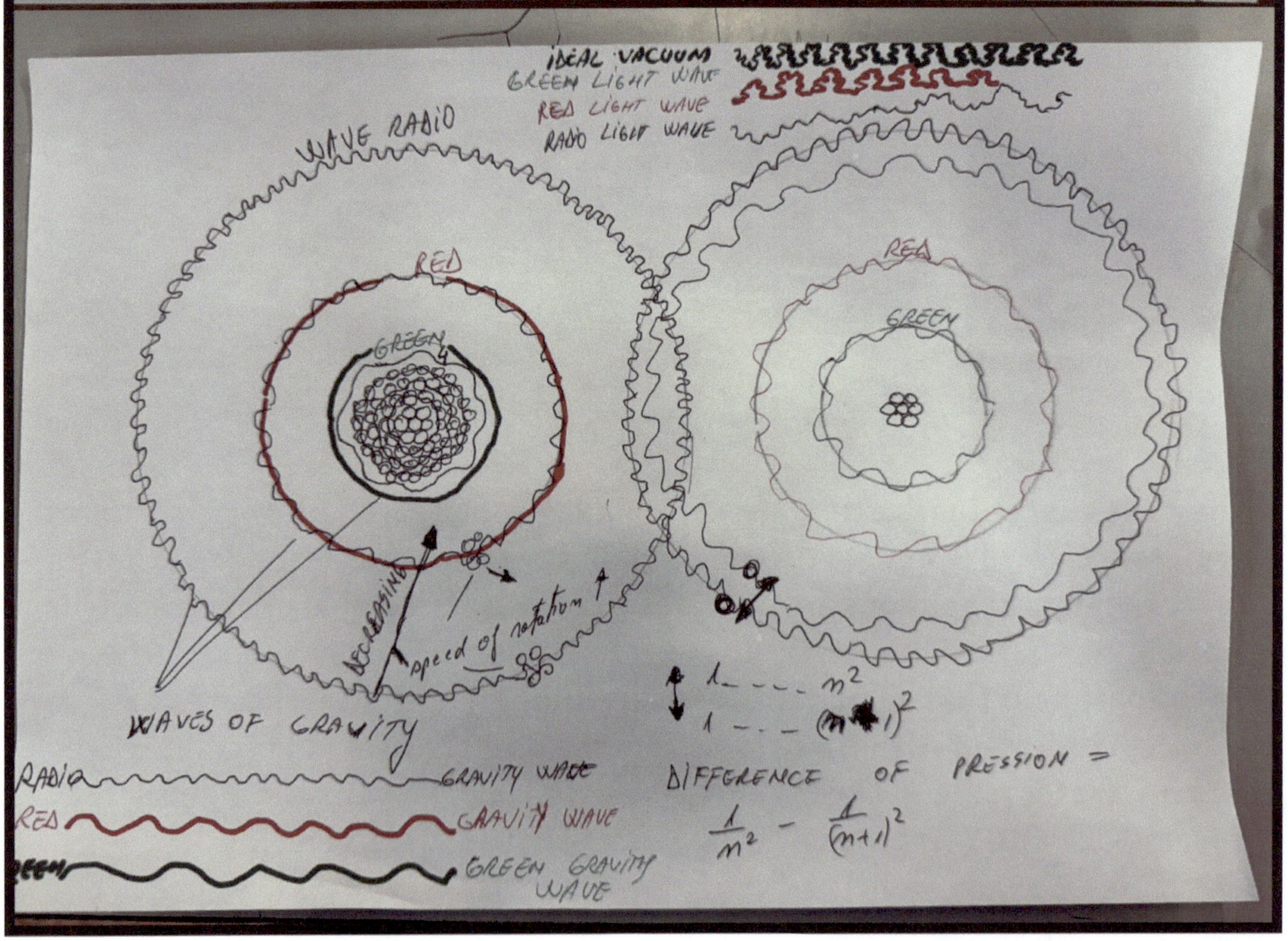
IDEAL VACUUM
GREEN LIGHT WAVE
RED LIGHT WAVE
RADIO LIGHT WAVE
WAVE RADIO
RED
GREEN
RED
GREEN
DECREASING
speed of rotation
WAVES OF GRAVITY
RADIO GRAVITY WAVE
RED GRAVITY WAVE
GREEN GREEN GRAVITY WAVE
1 ---- n^2
1 ---- $(n+1)^2$
DIFFERENCE OF PRESSION =
$\frac{1}{n^2} - \frac{1}{(n+1)^2}$

9.There can be no gravity without electromagnetism and vice versa.

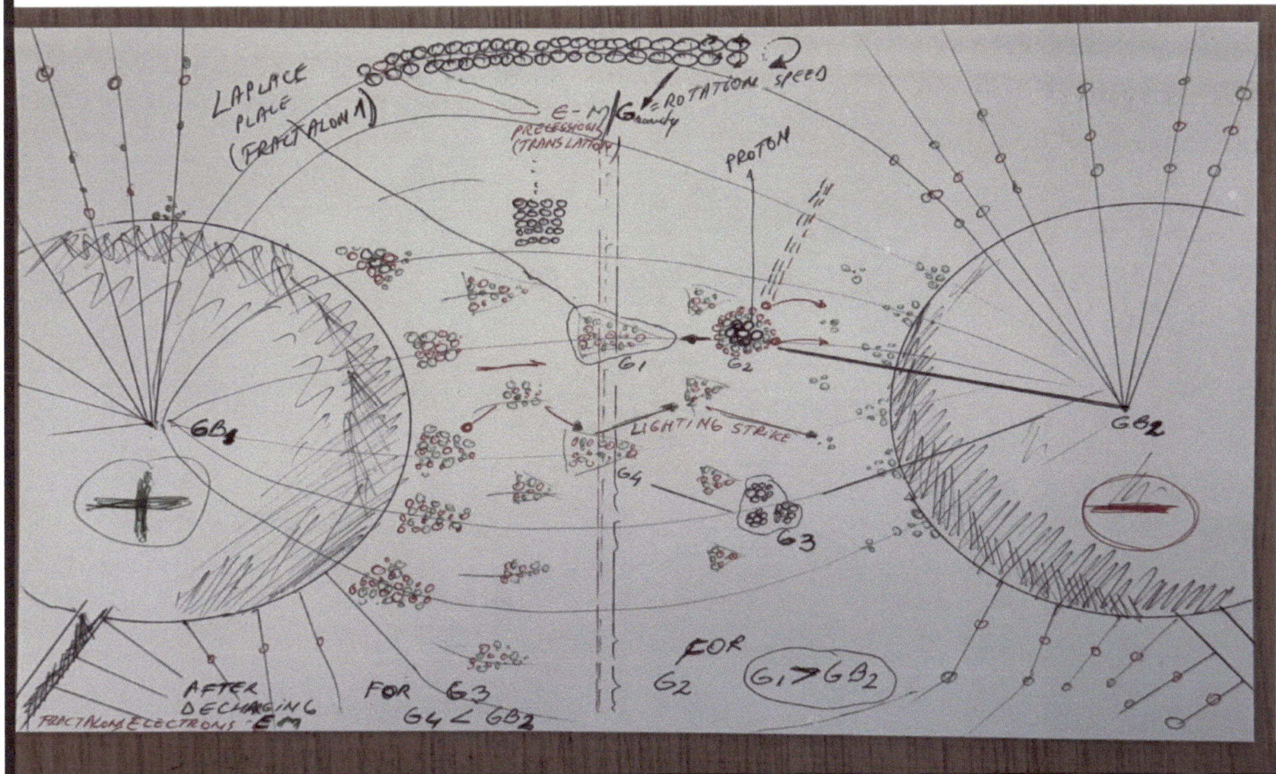

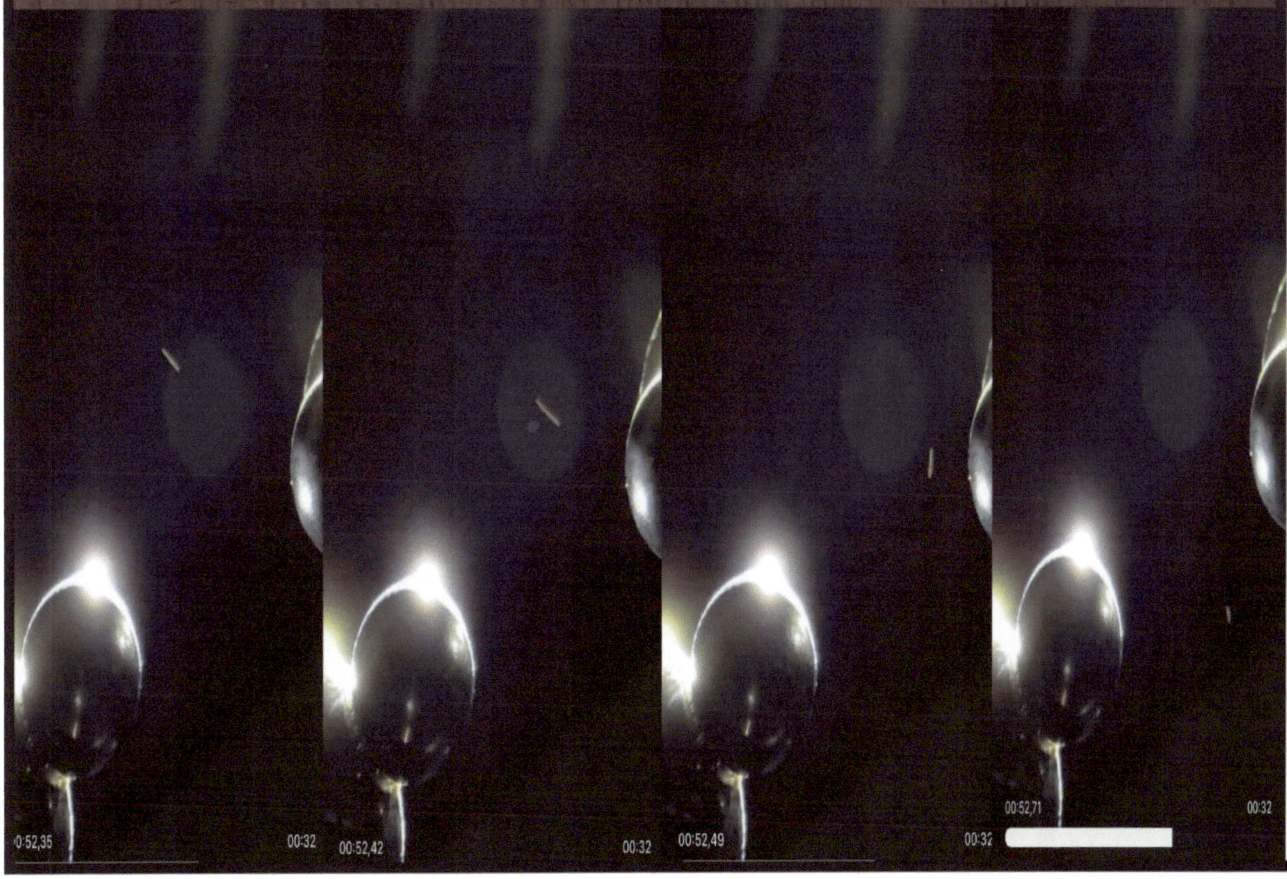

10.The frequency of gravity is the maximum frequency in the universe, and the speed of gravity is the maximum speed at any point in space, at that place light can have a speed identical to gravity or be reflected at the resonance limit with fractalons with a similar translation.Light travels, is contained, in the gravitational wave and its frequency will always be lower than the frequency of gravity.

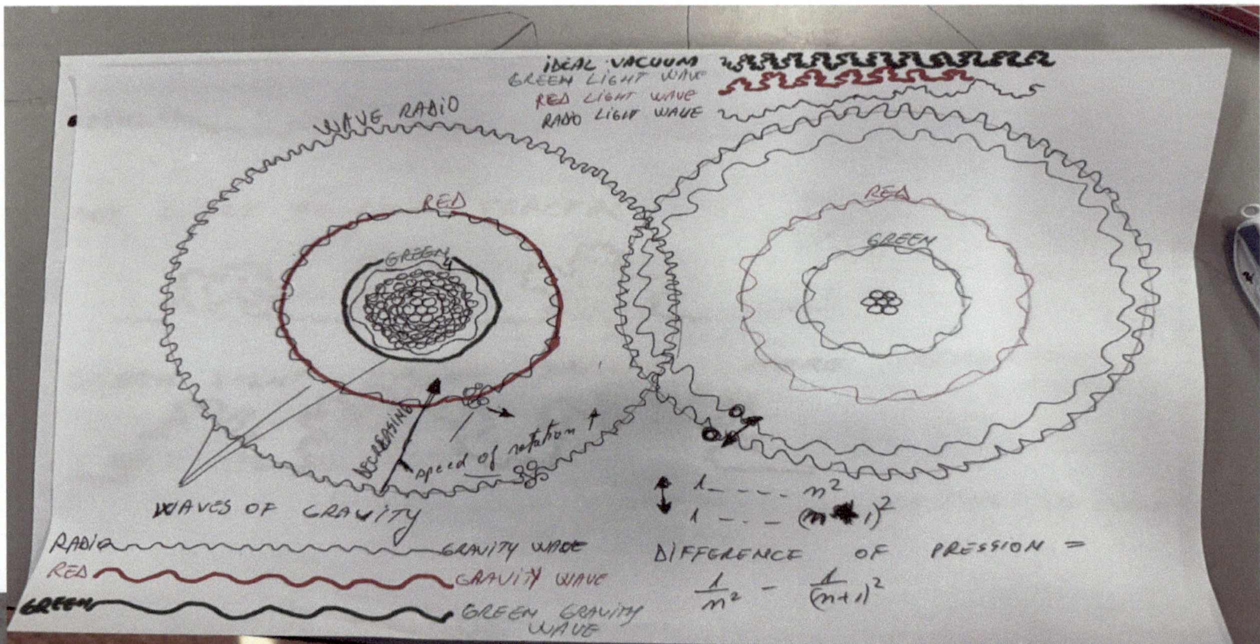

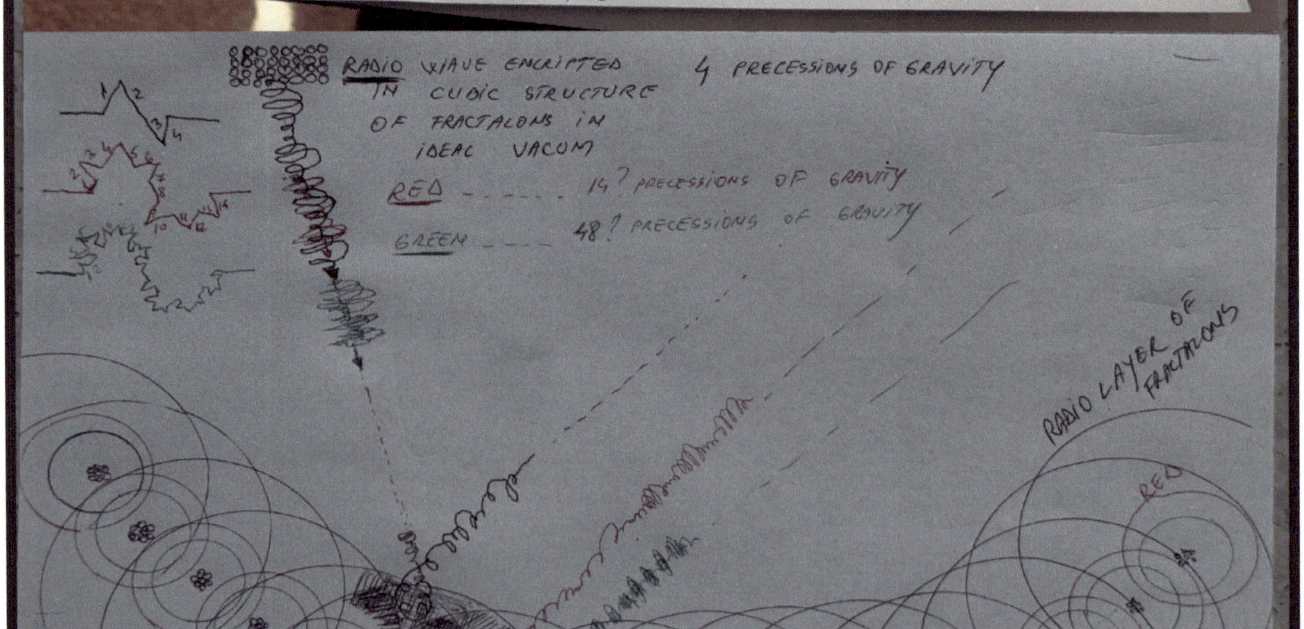

So as a direct consequence, the famous $E=mc^2$ becomes

$-E=nVg^2$

n=number of particles stuck in atoms

Vg=speed of GRAVITY=speed of contact rotation of a 3D surface always ² dependent

(nuclear reactions in next book)

The speed of gravity created by a group of atoms or an atom decreases with the square of the distance. The speed of light and the speed of gravity being always equal, we know from experiments that when approaching an atom (or a group of atoms with a common gravitational bubble) the waves in the visible spectrum will be slowed down (the refractive index of a material) or completely reflected, for the simple reason that the rotation speed of the fractalons that compose these layers decreases by losing equatorial contact (fractalon part 1). If they are reflected, the reflection will occur in an area where the fractalons on one side and the other of the Schrodinger bubble have different rotation speeds.Cubic structure is the most reflective.

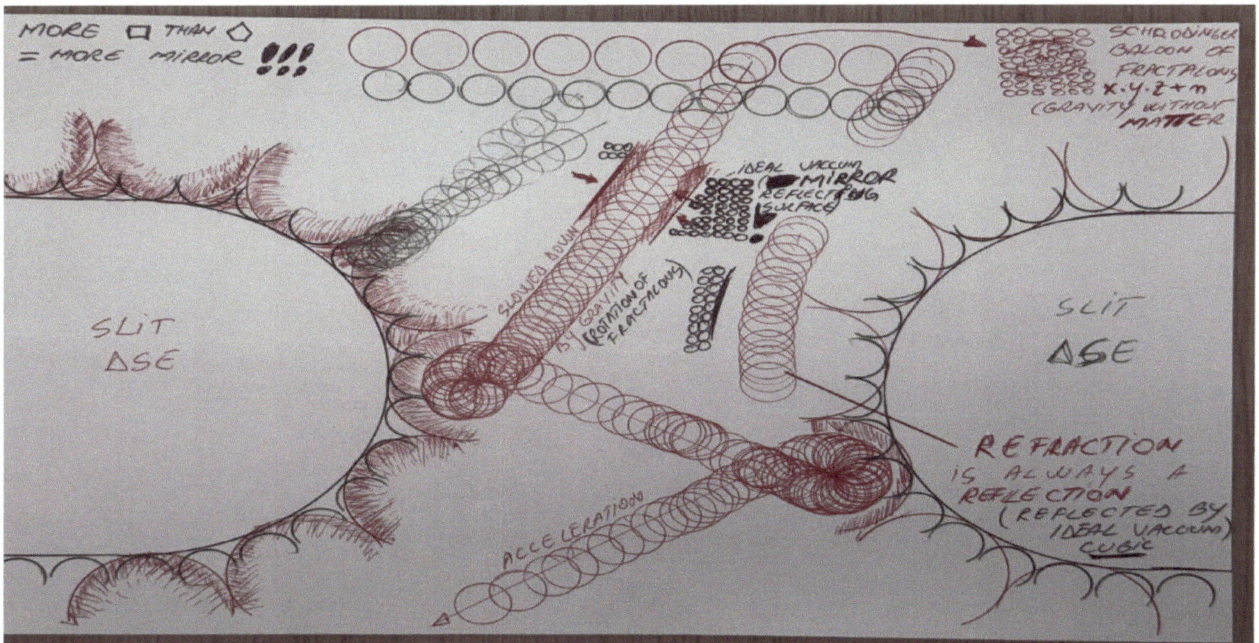

11.The number of additional particles added to the IDEAL VACUUM will define: 1- the amount of translational energy in the universe (gravity, electromagnetism, light, temperature) now also called dark energy and 2- the amount of matter (particles without individual rotation). This proportion is the only parameter of the current universe - THE UNIVERSAL COSMOLOGICAL CONSTANT OF THE UNIVERSE.

Cosmological constant : 1 for 8 or 1 for infinity.

Thoughts and reflections not thematically organized as I think, maybe they will help you understand better.

1.The speed of gravity created by a mass or an atom decreases with the square of the distance. If the two speeds are equal, and we know that when approaching atoms, waves in the visible spectrum are slowed down (the refractive index of a material) or completely reflected (depending on the angle specific to each atom or group of atoms). The simplest example is a glass prism or the surface of water. So the distance at which light travels from the atom (or a black hole) has an influence on its speed. The closer you get to the center of gravity, the more you slow down. And we already know that. So the closer light gets to a black hole or the center of atoms (protons and neutrons), the more the light and the speed of gravitation slow down with the square of the distance from the center of each atom (because gravity is created by each atom individually, and if the atom becomes giant like a black hole, then the light slows down directly with the square of the distance from the black hole).
We observers are made up of atoms, so the information that arrives is always slowed down. If we have two different observers both watching a giant supernova explosion, one looks directly at it and the other has a mirror in front of him. Only the waves that reach his detectors are described by each observer. The rays of the light spectrum are reflected by one of the two, for him it is equivalent that they did not exist, the gravitational waves reach the atoms of two, the two will affirm that the explosion with the gravitational wave took place.

2.Imagine a whale and its son. The whale is in the Arctic Ocean, and at night, the water has frozen around the whale and its son. Its son wants to leave. Another whale, far away in the ice, watches the scene like a God. What will it see? Initially, both are in the same ball of water surrounded by ice. The son tries to move away, but he can't move forward because of the ice. He waits for the ice to melt around him. Once the ice melts, he moves forward, and the ice re-forms between him and his mother. The speed at which the ice melts around us and then refreezes is the speed of light around us. If you can spin around x times and throw a very hot object into the icy ocean, like a glowing ball of metal, and the ice will reflect the heat infinitely back into the ball, the speed at which the ball melts the ocean as it moves forward and then refreezes is the speed of light in a vacuum. The ball is the PHOTON=The Schrodinger equation like baloon of fractalons with a number of suplementary fractalons relative to the cubic structure.

3.Gravity and electromagnetism are inseparable, electromagnetism always follows gravity. Electrons always remain in the space between matter and vacuum, they cannot travel alone - electrostatics .If they travel to release the pressure from the lower layers they will do so in the form of successive layer discharges in the form of electromagnetic radiation, radio waves, light, etc.Schrodinger ballons (encoded in gravity wave,changing cosmological constante for a entire group of fractalons from ideal vacuum) that is why in the Whimhurst machine we always have several successive flashes when we bring the metal balls closer, different layers of the peripheral fractal are discharged

using very fine particles we can visualize the fracture lines of the ether, the changing precession lines in the ether and where the lightning will strike before it occurs (video on fractalon.fr)

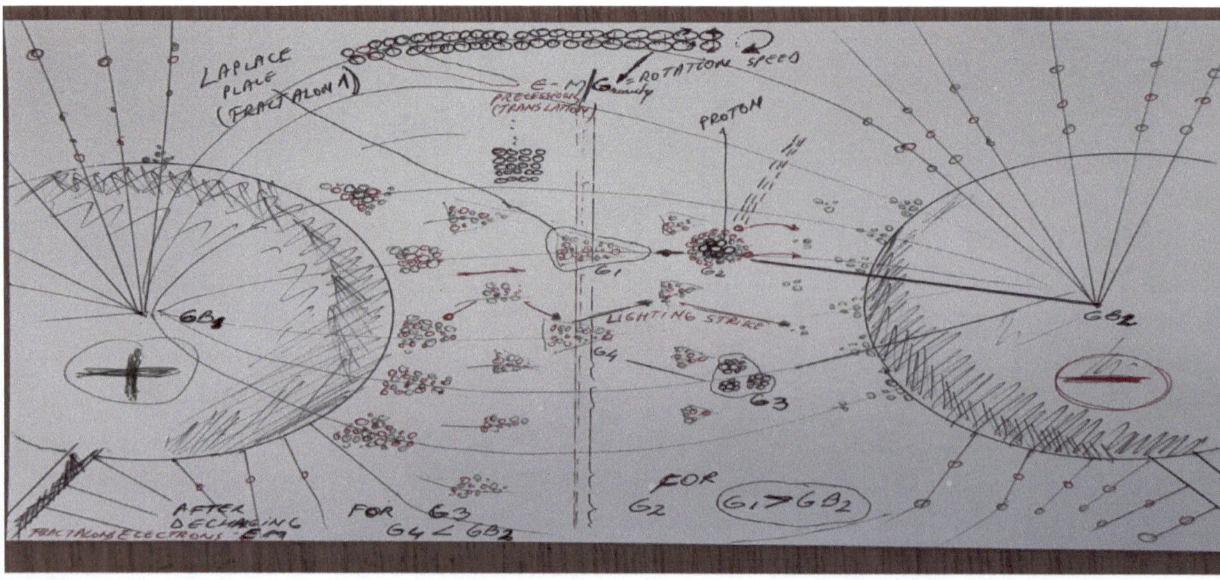

Light is produced and influenced in the resonance zones between
gravity and electromagnetism (quantum effect), when it passes .otherwise the pressure disturbance that it creates around it is enough for it to move
in a straight line,the Schrodinger ballon is reflected by ideal vacumm around him .Like the balls on the shore,in the beachwater, sometimes take by wave sometimes not, depending of the resonance of his gravity and gravity of the wave.
3.In physics size matters where you are located as in mandelbrot's fractal on the fracture and what size you are .Bubbles on the wall is not a force , is gravity and electromagnetism, dust changing direction is gravity and electromagnetism.ALL IS GRAVITY AND ELECTROMAGNETISM

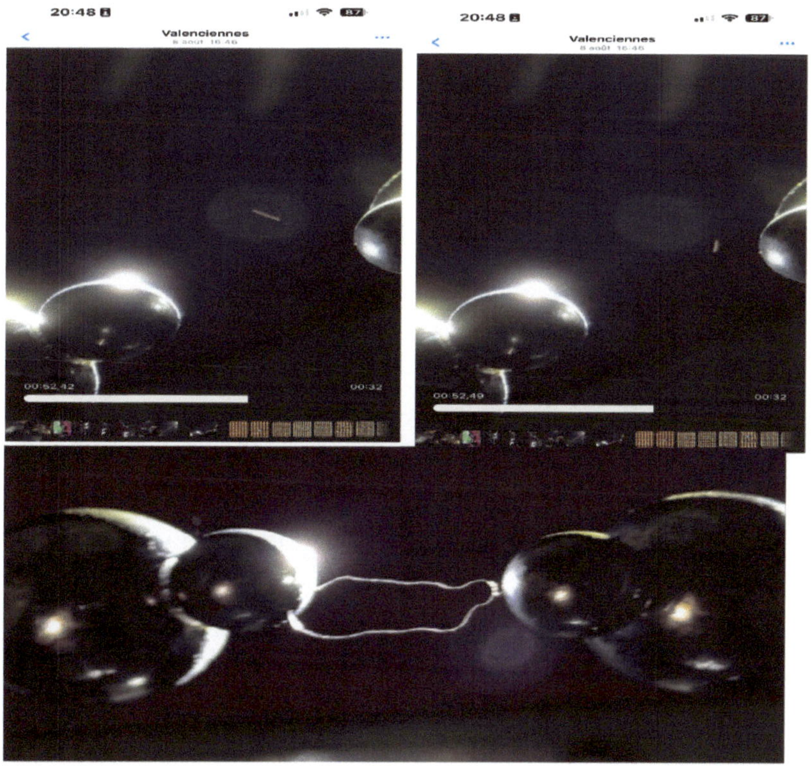

We see fracture of space we see the gravitational auto arrangement, the crystallization of space around atoms, the lighting is the path of electric light in this configuration, if you can see inside a copper wire you will see the same thing. The gravitational arrangement is the same thing gravitational layers of different substances, like the surface of the ocean, like oil blaster on the water, like chemical reactions. Is an arrangement of gravitational and electric geometry of atoms and electrons, you cannot separate GRAVITY AND ELECTROMAGNETISM. The thing of electrostatic energy arrangement of hair when you are electrified is only because you have a continuity between atoms between layers of electrons like in Mandelbrot periphery, every hair is a big fractal balloon on the big central head.

4.Double slit show you just holograms encrypted in cylindrical reflection, and Steve Mold has an excellent video about that, every time you take a picture in dynamic view all reflected light gives you when you zoom a hologram that is already in the picture, is not interference of the wave is just reflected light with similar comportement in motion. You can obtain different version of double slit, direct light or with shadow, the big spot is given by distance between cylinders (foot, inbox forks, straws,

and the little is given by the distance between cylinders or reflecting spheres the light (groups of atoms for every length wave forming a common bobble of gravity without matter like a soap bobble around 3 soap bobbles) and this virtual bobble had the radius of individual distance where the wave light is reflected or formed.At the different distance you obtain different superposition of different cylindrical mirrors, the same image is obtained virtually when you look a grilled paper focusing behind her (in the way you look at stereograms)

When they start quantum mechanics physics they start with wrong premises the things they see are interferences because they use light but they forget electromagnetic wave reflect on metal mirrors at different distances, depending on the wavelength (for that radio antennas are made from wires reflecting occur at certain distances from the wires and red light and green light are reflected at different distances from the mirror so after reflection we have a distance between rays of light that come together so instead of viewing double slit just like a simple Cylindrical mirror reflection they interpret this like interference. Same phenomenon can be reproduced in different circumstances and always is just reflected light on cylindrical mirrors. All wavelengths have the same speed in ideal vacuum and "transparent "vacuum because they are all encoded in gravity wave like in the photo. They will be reflected by an atom at the distance (in number of layers 1,2,3,4,5..) of resonance with the wavelength (the formula of spectral lines). But they will be reflected also by a group of atoms at the equivalent changing precession distance giving a resonance. And the reflection depends on of the organization of the vacuum surrounding matter. The gravitational and electromagnetic organization (precessions of Fractalons). Electromagnetism and gravity cannot be separated.

Electromagnetism is encoded in
gravitational waves.Like we see in the experiment with the very tiny particles that follows the gravitational (fractures),borders of the vacuum.Cristalisation follows this invisible organisation of the vacuum ,Chemical reactions follow this organization of the vacuum,the layers of different liquids (with different density)substances follow this organization of the vacuum and most important for the physicist the electrons and light in lighting follow this gravitational organization of the vacuum like in experiments in the video fractalon.fr.We can see where the lighting will strike before he strike,and if we deformed with finger we can see if he will touch the finger or not .Only the atoms with continuity of peripheric organization act like a electrostat ,the closed little spheres are not affected by Fractalons electrons passing.(photo) .The simplest way to imagine the fractures of gravity, the paths of changing precessions whit same frequency is to look at the image a and the video or like I say in Fractalon volume 1- a soap foam around matter .

5.The wavelength of different colors like of the sea in the photo doesn't show us the light is a wave ,the speed of the real wave is the speed of sound and the speed of gravity.Wave length of light is just encoded frequencies in the wave of gravity.And the sea wave is encoded frequencies in the speed of sound and gravity of the water.But the only two real waves in the photo are gravity and sound waves,and they are not visibles In the photo.

6.The wavelength of different colors like of the sea in the photo doesn't show us the light is a wave ,the speed of the real wave is the speed of sound and the speed of gravity.Wave length of light is just encoded frequencies in the wave of gravity.And the sea wave is encoded frequencies in the speed of sound and gravity of the water.But the only two real waves in the two photos are gravity and sound waves,and they are not visibles In the photo.
To understand Amper law like in the video fractalon.fr you apply my advices to solve the global equations and Fractalon unified theory (so
the balls pool)and you understand that :you must imagine electrons like I say in Fractalon first part book like boats ,imagine a lot of boats like a chain boats navigate at 10 meters one of these first in same ways and second in oposite direction ,what kind of water wave (the ether)they will cross and you understand why current in same direction attract the wires and why currents in oposite direction repeal the wires

One atom or one group of atoms cannot emiting in same time two different length wave ,it,s all about pression so you reflect or emit depending of the precession or rotation of neighbors (Fractalon part 3) depending of gravity (rotation)and electromagnetism(precession) but precessions gives us a region (a ballon ,a Scrodinger 3d balloon like in the video of Steve Mould with resonance of a ballon) whit similars frequency of precession and a fracture limit like a invisible soap ballon(Fractalon first part explanation).
So to understand this you need to understand that :in every sea wave or water wave you have just a ratio ,a equilibrium ratio between energy carry by gravity (rotation and precession of Fractalons with result the mass and electromagnetic-electrostatic organisation of water) and energy carry by sound waves rotation of atoms and molecules deforming space .
Similar in sound wave is a ratio between energy of gravity of air(gravitational waves with the speed of light) and sound waves rotation of molecules of air.
The frequency of seawaave don't give you the speed of light either the speed of sound.
The frequency of a sound we listen either.
The maximum speed of a sea wave or a sound wave is the speed of sound.
Similar for the light the maximum frequency she can rich is the frequency of gravity.When you heat the bottom of a bottle full with water the wave you will see in the top are formed with the speed of sound ,same thing like the electrons in the wires ,the wave of electrons in the wires travel at the speed of gravity so the speed of different lights ,is very interesting to sea in what wavelength energy travel in a wire.In air is very simple, is the lighting scattering but visible light wavelength are reflected by atoms so is impossible to travel in this wavelength .
Why a wave of the sea become bigger and bigger?Because, if the speed of the boat is inferior at the speed of sound in the water the boat will touch the precedent waves like in the photos .In the surelevated water in front of the boat we have encoded like Fourier transformations ,like a fractal in time all precedent waves.
So the wave of a boat or a wave in generally is a fractal in time depending always of the differences between times of rotation and time of translations of participants relative to neighboring.

So finally to answer in Fractalon modern new quantum physics way at the question why the oscilloscope in Veritasium AlphaPhoenix etc show this curve of current is very simple .In a wire and arround the wire ,electric current, so electrons will travel in different positions of a cylinder like water in a tube ,and the radius of this cilinder dépend of the tension of current. electrons have precessions,they need space.The vacuum have space but if electrons perturb in concentric waves the vacuum they will create precessions inthe peripheric layers of vacuum at the
speed of gravity and light in the vacuum ,we can say the tension .The pression of the vacuum is maximum relative to other electrons arround mater or in Light .Imagine a lot of fish in a ocean starting to freeze from the bottom(the atoms) they will jump but they are not attracted back by the water and ice ,they are repealed by the air,the air have space for them but more precessions than they can create so the are gravitational push back ,the fish trying to jump are electrons.So every time the front line of traveling lightning is a sphere So the Electrons in and arround the wire travel encoded in different wavelength (remember the sea waves) but the waves in different layers depending of the organisation of the atoms, like the asteroids of different size traveling in solar sistem ,travel at different distances of atoms or group of atoms ,so at different speed of gravity.So in the photo the oscilloscope showing the beginning the electrons moved by the wavelength at maximum distance relative to the atoms or group of atoms of the sistem.But the majority of electrons don't travel encoded in this wavelength. Why I say moved or encoded ?

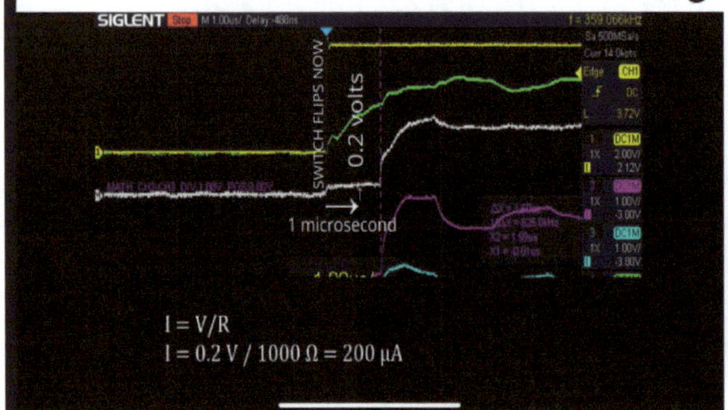

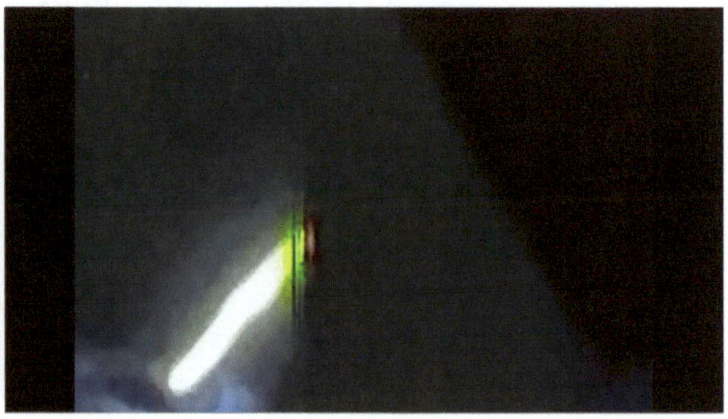

Because in order to travel the front light of a wavelength must be REFLECTED and this is the key of all this book and to be reflected the frontline must crossing a atom
or a group of atoms to create a fracture of space a surface of reflection at a certain distance from the atoms or groups of atoms if not we just have a emission of radio waves light etc.To understand reflection the simplest analogy is to imagine a comet in two stars system.

Only precession is reflected, not rotation. Rotation is only slowed down or accelerated. And gravity depends on the rotation of fractalons.

The amplitude of a wavelength, the size of bubble of pressure, and the frequency of light, every EM wave, like in the comparison with waves of sea in front of the boat where is a relationship between energy carried by precession of molecules of water (sound waves) and the gravity of the wave (a front of equilibrium between the two energies), in the case of light is an equilibrium between the energy carried by precession (EM, Light, electrons) and the energy carried by gravity (every Fractalon of universe) the mother of all waves are gravity, all other waves (light, sound, sea waves) have Fourier fractals in times of gravity created by deformation of space.

In electrostatic discharge electrons are discharged in layers, Fourier discharge, inverse-fractal in time, different shape and different light emitted because if we put a glass on electrode we change completely the layers and the fracture lines for discharge, they convert like the river to big fracture of metal (Ampere law again), in the opposite way you force electrons from the river to climb the mountain in the littles fractures.

7.Double slit experiment is just reflecting light on 3 cylinders mirrors(4 mirrors)

We can reproduce this experiment I'm many ways with different source of light and cylinders. Effectivement in a source of light we already have a density of light arranged like in the photo of a lighting,the density depends on the wavelength and how his light is already collimated by

the mirror atoms like in the photo so the final source of light is like that (photo)this image will be reflected in every wire 1,2,3,4,5,....like in the photos giving us the final image.Is not that thing like interference.To understand you need to see like a observer. In every black point of a double slit experiment or interference images you don't see a light directly or reflected in the mirrors,you just don't see in straight line any light. I made thousands of photos of " interference ".All you see is similar comportment in dynamic zoom or direct photo at a certain distance of reflected light (because the different wavelengths are reflected at différentes radius of mirrors)confirming Fractalon Unified Theory.At different angles you will see always same colors .To understand take photos of something in antimosquito patterns. Half of information of the photo is reflected light.But in all photos you have this.If you change the objectives,the distances,the zoom the interference pattern of same scene don't change,because don't depend of the mirrors ,or the iPhone ,or the screen,you can use the screen of the iPhone to zoom and search the "interference "pattern for different distances but all you see is just reflected light with similar comportment.This information is already encoded in the color of every pixel of the photo.Look at this examples to understand.

You can reproduce DSE with mirrors in front of the mirror (just put something opaque in front of every cylinder to block direct light reflection and keep only multiple reflection light from the sides)what you obtain is same curving pattern if you incline the wires of DSE,and you have a complete cercle of "interferences",in fact just reflected light in a cylindrical mirror at different radius depending of the wavelength and density of rays on each mm2.To reproduce use multiples sources of light like to understand how it works reflecting lights(holes is paper).To obtain them at atomic scale use iPhone to take photos of lighting in electrostatic machine. The space must be compress in equal volumes to obtain different lengthwaves,in RAINBOW orders.But the light is like resort,In same hole you can pass more green(little resorts)than red resorts)(Schrödinger 3 d balloon resonating)

Don't forget,in DSE half of the image turn in one direction and half in other but always perpendicular on cylindrical mirror (the wires)in fact are just multiples of 2 direct light,multiples of 4 reflected one time light,multiples of 3 reflected one time light ,multiples of 4 reflected 2 times etc.

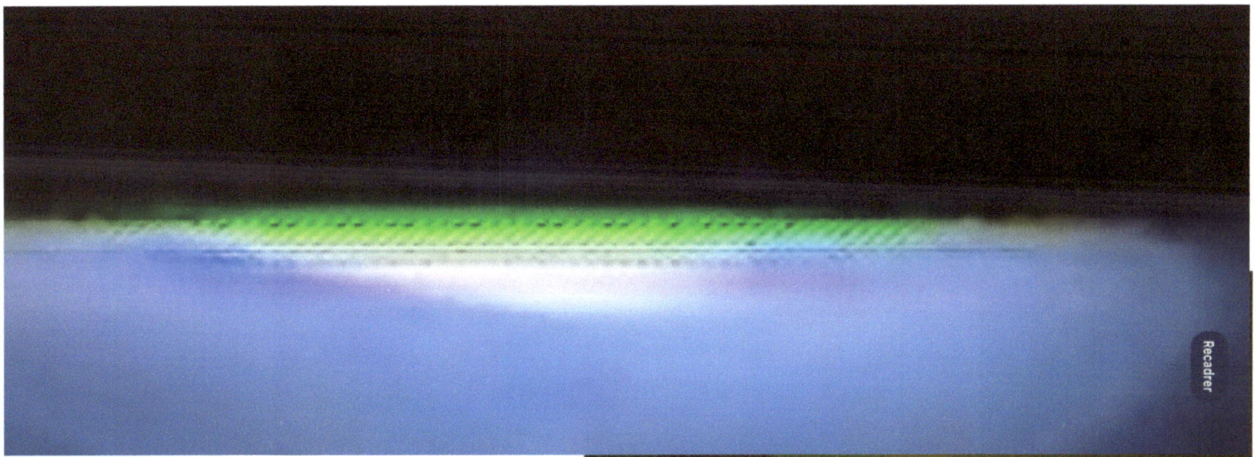
Recadrer

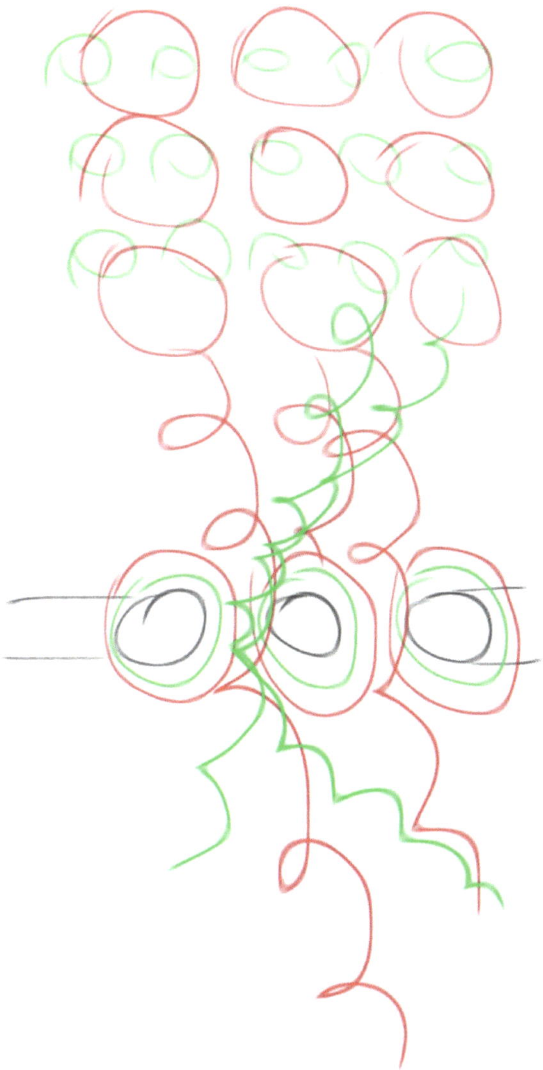

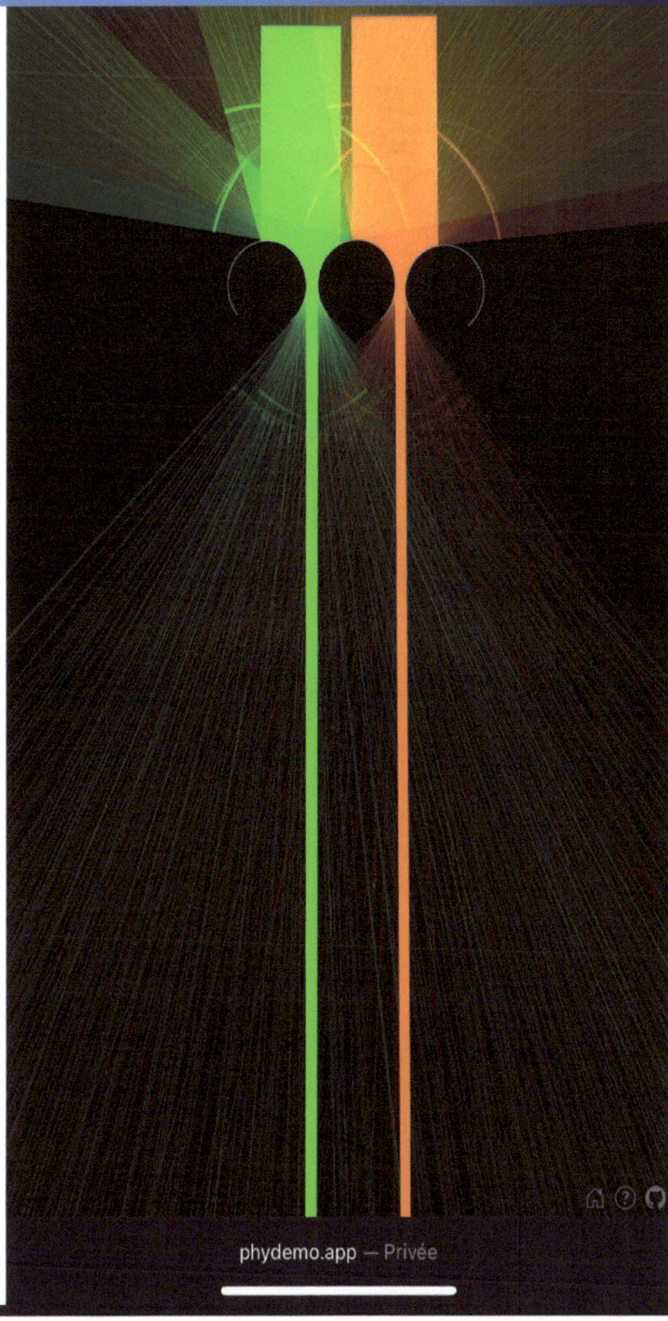
phydemo.app — Privée

After multiples crossing like
in the photo you have only almost parallel light and at this distance if you do the right thing with parallel wires you have" interference",if you do like me with intentional twisted wire you see all the reflection not supperposed, and is identically if you use big mirrors like forks or little mirrors like wires in DSE. But don't forget radio waves reflect at one distance from the atom,red at other distance ,and green at another ,and reflection is like that! not like In our books(refraction and other stupid things from actual optic physics are just particularlarity of light changing direction (I keep the word reflection)) at a resonance wavelength with gravity wave(wavelength of light is encoded in gravity wave ,for that all EM waves have some speed ,the speed of gravity,but the speed of gravity diminishes approaching individual or groups of atoms ,so the speed of light change.The perfect analogy is with a asteroid entering in our solar system ,if he is big enough the only thing to make him go back is the SUN,if he is little a planet whiteout atmosphere can do same thing.The reflection occurs at the limit of different layers of same gravity,you can see a little dust particle doing that ,the gravitational shape of the layers change approaching matter ,so this little particle of dust deforme the space of the Fractalons in same way that a light(or radio waves)with same size of peripheric limits ,the both are reflected at same distance .Charging atoms with electrons modify the limits ,the fractures of peripheric gravity, plus electromagnetism(speed of rotation and frequency of precessions)of vacuum.The molecules of gases,dust ,crystallize in this new fractures of vacuum ,for that metal lighting or glass lighting are different,and we discharge only one by succesives layers of electrons.
Like a fractal.A series of lighting change the configuration of fractures of vacuum,is like to have 100 balloons arround a body ,everyone with gas and you pop them one by one .For that the colour of the light is different in lighting,you need to discharge the same layers of all atoms in contact charged with supplementary Fractalons, the layers that emit the series of different spectral (Lyman ,etc...)and only if you have a fracture continuos of the gravity at this distance from the atoms or groups of atoms you can move the electrons in this layer.To do a analogy: to move our moon from earth to Jupiter you need to put at certain distance 1000 planets quasi identical in size with the earth between us and Jupiter, at precisely distances,so the moon will follow this path.

8.All what actual physics call now "Interference",interferometer,prisme refraction,mirrors interference , with the conclusion "EM waves are waves, acting like wave based on DSE ",all that are just delayed waves after a reflection (in fact changing direction at the limit of two different gravity layers)reflexion is an particularly case :the wave go back like a steroid or comet arround sun.Every change depending of radius of lenghwaves. The only wave that are never delayed in univers is gravity ,but she can be diminished in speed.All others wave are encoded in different waves of gravity, they jump from one wave of gravity to another, from one layer to another.

The light is for gravity what this sea wave is for sound and gravity waves.Sound waves are Fourier fractal transformation of the space (time of rotation and translation changed in x successive rotations of Fractalons).In a sea wave we have encoded a(the eye is a oscilloscope for this wave,like the oscilloscope for the electric current)RATIO of equilibrium between the gravity of a volume of water and energy carry by multiples precessions of molecules of water in sound waves (but all works only if is a continuity between new big deformations of the space).Similarly for the light is a RATIO of EQUILIBRIUM encoded in gravity wave between energy of translation generated by gravity in every point of the space depending of speed of rotation of fractalons(Fractalon part 3)and energy of precessions carry in the Schrodinger ballon of a region of Fractalons (the light) but this region I think ,I'm not 100% sure, can exist only if the density of Fractalons per unity of volume increases with 1for xyz.The sea waves if are not reflected by a rock ,if is a sand beach, they came with water.The light realy come with a suplementarry fractalons(n for xyz)

Every photo ,every image we see,with our oscilloscope (the eye) is a reflected image ,in every image a part of the light we see is reflected light :or in the source like in lighting (directly on emitter,multiple reflection between atoms) or in the detectors or lens .And different wavelengths reflect at different distances from the atoms or groupes of atoms depending of their energy

So in DSE we see just a sum of reflected image ,of packaging of different light (every wavelength has his own pattern encoded inside)on spherical or cylindrical atoms or groups of atoms . CYLINDRICAL MIRRORS TURN THE IMAGE AT 45 degrees and two images combined gives us this ("interference" pattern)You can obtain same thing with football balls and two big cylindrical metals.And a hole.Or after you eat a lot of cherish with the pit.You can call it "Single split in the black hole of idiots" "and image of the "interferences" of all idiots
physicists "
In every line of reflected line of Z the smallest spot have same dimensions, change the place but no dimension,because at atomic scale is not more cylindrical line is just 2 or 3 atoms and the space between them ,the smallest spot have always Same dimensions after multiple reflection on cylindrical mirrors the dimensions of lengtwave of red or green light.IF YOU KEEP THE WIRES IN YOU HAND ,or the splits,and you twisted or you turn them slowly you will see every " pixel" every wave length of green or red moving independently (you see un second phenomenon,the green light split in a card in red light:) but explication of this ,other time).You don't see any " interference" ,you can follow in the wall every independent wave of green or red light,and if you are obsessed you can number all the independent waves per nanosecond in the photo of emited light ,and you will see in the wall you have a ratio of this number of pixels.At atomic level ,the form of cylindrical mirror in DSE is like in the beggining of this book(the magnets).And the radius is different for the red wavelenght and green
light.And reflection is all or nothing so, the angle change,but the information ,the energy is the same (a invisible ball of Fractalons at the speed of gravity ,with desorganisation and reorganization creating this Schrodinger like equation image of light.

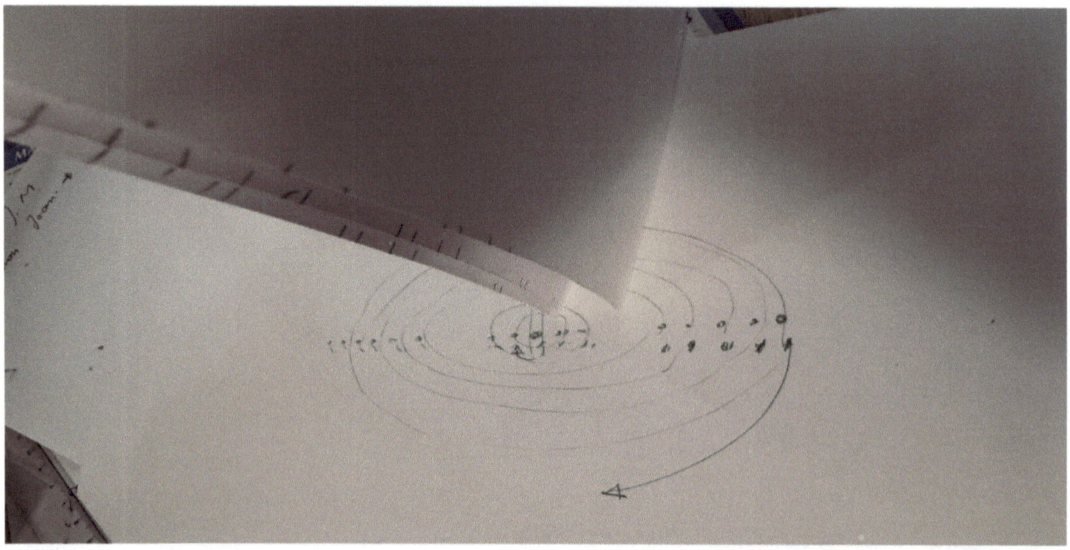

So to be clear in DSE we don't have 2 new secondary source of light like in all physics books with interference showing the wave nature of light,we have 2 giants panoramique projectors like 2 cinema projectors,projecting a image full of red dot or green dots with size of the length wave of green or red. If you turn the slits ,or the wires,360 degrees,and take a image for every angle for a certain distance,and after move 1 mm the screen and repeat again and again you obtain a tomography,a CT 3d image of reflecting image on 2 cylindrical mirrors,and after that when you do slice in different plane you can see every independent direction of every independent ray of light emitted. DSE SHOW YOU THE LENGHT WAVE OF RED AND GREEN LIGHT,and light propagated in straight line in air and is curved in different layers of different gravity ,don't show you the light is a wave.Analogy :the oscilloscope don't show you the EM waves in current are waves ,just a frequency of changing in this wave ,you don't see frequency of light of EM waves that carry the current (the electric current in electric lines and arround them is carried encrypted in different frequencies of EM waves,in the wire and arround at a distance from the wire in visible spectre FOR THAT IN LIGHTING we see light,the electrons loosing their translation energy in different layers of gravity arround atoms or groups of atoms,and inside the lines different frequency from visible,infrarouge until gravity frequency,THE MAXIMUM FREQUENCY IN UNIVERSE-electronic microscope working in this principle)like in the frequency of a sound you don't see sound waves.You can use this frequency to measure the speed of real wave but the wave you see is not the real wave: THE GRAVITY

In the photo with laser in bank card we see it real interference between waves ,in order to interfere different waves is logic they must change frequency,so they change wave length ,green become red ,this I a real interference pattern ,the color change not the pattern we see,the pattern in DSE is just reflection.

In a prisme what you see is changing speed of gravity of the layers where the different wavelengths of light travel ,not changing speed of light ,just red travel more in this new gravity environment.,the difference is only on reflection at the fracture at the limit of separation between layers. Yes the light is encoded in gravity waves but not all the length waves can travel in new conditions,some are reflected some are split just gravity pass

If you do the experiment with the electrons the results will be the same because deviation,the reflection is produced in resonance with gravity. All DSE is a gravity changing ,no matter with what you will do it ,is like a asteroid deviated or reflected by the sun,the dust in the photos and in the my chanel video is deviated by layers of Fractalons with peripheric precession identically with layer of gravity where are reflected if you decharge this dust of his electrons he will be reflected at other distance. The distance where EM waves or electrons are reflected depend of the gravity (speed of rotation of suplementarry electron charging the atoms in LAPLACE REGIONSlayers created by the material of DSE.

We can obtain higher resolutions than the wavelength? the answer is yes! Because the resolution of the iPhone sensor depends on the density of atoms that capture electrons, which can in theory, just like in the electron microscope, approach the size of the electron fractalon, only that we recover the information in x-rays or gamma rays, in our case the sensor can capture green wavelengths over time in different positions and super-positioned we obtain resolutions much higher than the wavelength, on the other hand, at the same moment the laser cannot emit a higher resolution image at the wavelength of green light, when I say the same moment I refer to a complete tour of the fractalons that will emit the green frequency, which depends on the rotation time relative to the hypothetical fractalons in the ideal vacuum (Fractalon part 1), that is, the speed of gravity in the laser that will define the speed of light at the atomic level where it is produced, it will then accelerate in the air to the speed of rotation of fractalons in the air. Electromagnetic waves as well as electrostatic or electronic fractalons (ampere" force") are always reflected by the strongest vacuum and then in this order: the more a medium has fractalons organized in the cubic structure the more it will reflect than one with fractalons in the intermediate structure and in the hexagonal structure there is no more rotation and individual precession. Because fractalons in the ideal vacuum are able to produce more precession having a higher rotation speed=pressure. For this reason, electron or electrostatic fractalons cannot leave matter, an electron cannot leave for the moon, it remains in the electromagnetic field of the earth. (maybe in the form of light he can !?but in fact it only pushes another one, it does not really leave like in the electric current).

You cannot create one photon this is a lie ,you create one "wave " of red in a concentric place,one Schrodinger baloon with "n" suplementary fractalons for xyz of IDEAL VACUUM space ,all arround just one single concentric waves ,if you create just one you will never seen ,and in fact is absurd you cannot fracture the space to do one photon of red ,you need all the space for the entire length wave of red light.

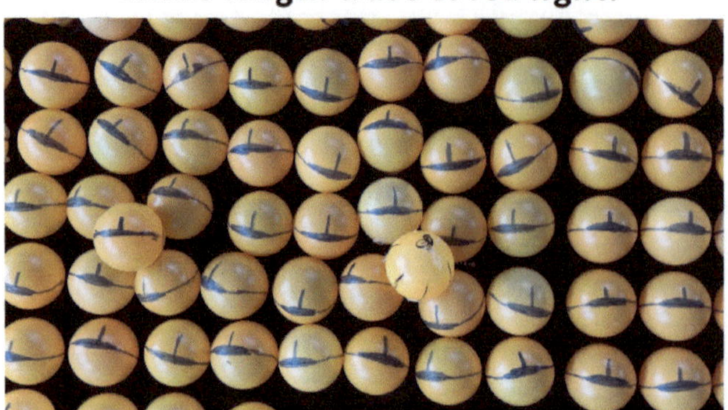

FRACTALON UNIFIED THEORY SUMMARY

Newest Journal of New Physics:
Interview

Newest
Journal of new physics

Unified Theory, How Higgs's boson creates Gravity and Electromagnetism, how atoms are created?

Interview with Dr Terec Traian, author of Unified Theory, March 2025

fractalon.fr

Fractalon Part 1

Dr. Terec Traian,
Amazon ISBN **978-2958148201**

Fractalon Part 2 Traian Table

Dr Terec Traian,

Amazon ISBN **978-2958148232**

Fractalon ,The 4 "Forces"

Dr Terec Traian
Amazon ISBN **978-2958148249**

How the Higgs's boson(i.e.fractalon)turns into a lepton, like a "Swiss Watch Tourbillon"
All the energy of the universe is stored in a network of rotating Higgs bosons in a nearly cubic structure, "right" because each one has its own space, no one touches it and each spin undisturbed. This rotation represents the speed of light.
When you put one more here and there, it will be 8+1 odd neighbors ,4 rotating in one direction and 4 in opposite direction, new one it will deform the network, it will start to spin ,like a tourbillon of a Swiss watch, it slows everyone down, it is no longer Higgs boson, it is now "left" lepton, it steals rotational energy and transform it into translational energy (precessions and translations).
Two disturbances appear because of this: GRAVITY and ELECTROMAGNETISM.
When 6+1 left "communists" leptons are slowed down enough they get stuck in a PROTON (7=2+2+3=u+u+d)
They do not spin individually anymore, they just warp the clock and try to jam it, they turn more Higgs bosons around into electrons and slow down the rotation of all even more.
The whole Universe I like a giant SPACE-TIME clock.

I.R: Dr Terec, can you explain to us in a few simple words this theory and how it differs from other theories, such as string theory?

Dr Terec:

*Of course, first of all, it is the only theory that followed Eistein's vision that defined gravity as a deformation of SPACE-TIME.I have taken this concept further and present you with an extremely logical explanation of Gravity, but also of the other three fundamental interactions(electromagnetism, weak force and strong force)as well as perturbations of SPACE-TIME. The entire universe is a kind of fluid clock in constant change with only two parameters: **The New Cosmological Constant** to define the deformation and **The Time** to define the number of iterations that will define its shape at each moment.*

*Secondly, it is the simplest theory through which any person on Earth without higher education in Physics can understand the Theory of Relativity, can understand how the Higgs's boson(i.e. **fractalon**) creates **Gravity** and **Electromagnetism**, **leptons** and **electrons**, **protons** and **atoms** and why the **speed of light** can never be exceeded.*

I.R: If it's so simple to understand then why haven't I understood anything until now, can you help me?

Dr. Terec

(Laughs ironically) Of course, now please use your imagination for 3 minutes. Try to imagine somewhere in space, like ISS, a completely empty swimming pool and start filling it with balls like in the picture below and throw as many balls as possible into it ,until there are enough to form a perfectly cubic network. Regardless of which direction you threw them in and how hard they were spinning at the beginning, all the balls will self-arrange in a perfectly cubic network, in columns of balls that will spin half in one direction and half in the other.

1.First phase

2. Peripheric Motors columns for rotating the balls in the pool (every column rotate in opposite direction)

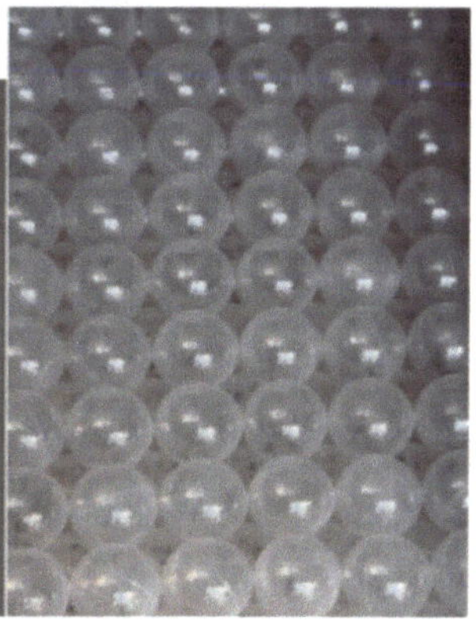

3. Self-arrangement in a cubic structure when they start to spin

*And now comes the culminating moment of the whole theory. The next 2 balls that we will introduce will explain 3/4 of the entire universe. When in this pool * of 20/20/20 balls that spin barely touching each other and we call them Higgs Bosons or Fractalons ,we will introduce a **single new** additional ball, this new ball will move differently and will force the others to rotate differently as I explain in the Fractalon short videos on **youtube** and in the 3rd part in my book Fractalon: The Forces.*

**(which I reproduced, in the middle picture shows some of the columns used to start Higgs bosons rotating half in one direction and half in the other, right picture self-arrangement of the balls)*

Two disturbances will appear in the pool, the first one is a concentric decrease in the rotation speeds towards the additional ball due to the loss of equatorial contact and the appearance of sinusoidal contacts that will generate precession, and the second phenomenon is the appearance of two areas with different rotation and translation around the additional ball which will receive a double 8 rotation, half-spin, 720 degrees, and we can already call it a lepton. As we get to introduce other additional balls(leptons) into the pool the rotation speed will decrease more and more, more translations will appear, more precessions, that will deform the cubic structure.

When the deformation is strong enough and the rotation speed is low enough, the additional balls will change places with "Higgs bosons" still organized in the cubic structure, they will steal a little of each one's rotational energy and will move towards the center of the pool due to the different rotational speed of the 8 neighbors

*We know this phenomenon as Gravity. The impact between the areas with different rotation and translation will cause some of the precessions with cyclic frequency given by upside-down rotation of the leptons to propagate in the network, what we now call **THE LIGHT**.*

*Both **Light** and **Gravity** are precessions transmitted from one ball to another. The transmission speed will depend in the speed of production of a precession, which is determined by the rotation speed of each "Higgs boson" in the system. For this reason, the **Speed of light** and **Gravity** are the same and represent the rotation speed of each ball (fractalon=Higgs boson). To change their position between them, multiple successive rotations are needed, so the translation will never be done at a speed higher than rotation, it being produced through rotations.*

We will obtain a cloud of leptons, of electrons. The space in which each lepton can rotate in double 8 between the 8 neighbors, this prison we can already call the weak force, that is, it represents a freedom of movement, a freedom of coupling. As we introduce more balls, eventually some balls will lock together and will no longer rotate individually. The first seed of matter appears like a snow flake in the plane in the middle of the pool .

7 balls are stuck together, stop rotating individually. This is the Proton.

a.one Lepton, deformation of the Network, Gravity and E-M without matter

b.2 leptons, Gravity and Electromagnetism, they change places between them, in the end the 2 additional balls are visualized in the middle.

c.1 PROTON 7 balls no longer spin individually, they just oscillated together

Zoom of the PROTON, Black dots are blocked balls, the red ones with increased translation, the green ones with high rotations, the red circles is the wave of translations, frequency of light.
Red and green dot balls are Electrons (the 2 poles x/y spin. The ways in which they change positions between them are orbitals.

I.R. So Einstein's famous equation $E=mc^2$ remains valid

Dr Terec

You see, all the initial energy is rotational energy, when 7 fractalons are locked in a Proton, they have a "remains" energy of oscillations(group precession) and a translation energy if the Proton is in movement, but majority of energy is lost in light, precessions transmitted in the network with the frequency of the every electron precession.

So, Einstein's equation remain valid with the specification that how many Higgs's bosons-Fractalons will be locked in matter, so much rotational energy can be recovered, so in equation $E=mc^2$, "m" represents the number of Higgs bosons that no longer spin individually, like those in a vacuum and "c" the initial rotation of everyone.

Per unit volume in a chemical or nuclear reaction, in order to release energy, the density of fractalons per unit of volume(electrons and complexes protons-neutrons=atoms), this density is the ***Cosmological Constant,*** *must increase, and vice-versa to retransform an Iron atom into Hydrogens atoms and then into vacuum, this amount of energy must be delivered to it.* ***It is easier to produce energy and matter than to produce a vacuum from matter.***

I.R Do you think the scientific community will validate the theory? You can validate it some way?

Dr Terec : *The physicist community had a harder time at first, but the chemists will do it right away. Let me explain.*

I called this pool of atoms formation FRACTALPOOLMETER and I filed a patent in France for the simulator, with the addition of new and new balls, the entire Mendeleev table was self-created as by magic, as in the crystallization of SNOWFLAKES, without any errors, in the order of the series an periods according to an exact formula :

2s+2s+6p+2s+6p+2s+8d+2s'+6p+2s+8d+2s'+6p+2s+14f+8d+2d'+6p+2s+14f+8d+2s'+6p+2s+20t+14f+8d+2s'+6p+2s+6p+20t+14f+8d+2s'+6p=222 TRAIANUM NOBLE GAS ATOM.

Why this exact formula? you see in the pool and in the photo of my book (in the picture below). Mendeleev become my new best friend; I love him! He is my savior, he is my code, and my validation of the theory.

Rule of self-arrangement of Higgs's bosons in atoms

A formula not discovered by the entire community of physicists and chemists because they did not realize that in the 10 "d" electrons 2 will become "s" surrounded by 8 "d" as all the reactions and properties of chemists have predicted. I have made you some reproductions of the atoms that appear. In the 3 images: on the left you have a proton, as we have already seen, the Hydrogen atom, in the center you have 4 planes of protons that regroup in a "reaction" into a Helium atom, and on the right you have the Gold atom. It is beautiful, isn't it?

Proton

Hélium Atom

Gold Atom

I.R. :How can the new theory change the world?

Dr Terec *: There is a latent energy between SOLID-LIQUID-GAS-VACUUM and vice-versa.Matter no longer has any rotation energy except for the electrons and Higgs bosons that will be immobilized in new atoms(in current physics they define this concept by* ***POTENTIAL*** *-just a nonsense word for physics).as the new theory creates the possibility of extracting latent heat between vacuum and matter (for example* ***sonoluminescence*** *already exploits this reaction of compressing the void in the matter).The major risk is that another country will understand the "****potential****" of the new theory and exploit this new technology before us.*

I.R.: Thank You and good luck Mr. Terec.

Dr Terec : Me, the STALKER ,thank you for welcoming me to the zone.

To understand Gravity and Electromagnetism we need to start from a system of particles(ideal dodecahedrons) without Gravity and Electromagnetism,just simples particles,like Higgs 's bosons.They will create Gravity and electromagnetism without our help just by rotations, precessions and translations.
In a system whitout Gravity or E-M you start to throw particles.

They will self-arrange into a cubic structure(ideal vacuum) through successive rotations and translations if the particle density reaches a certain level

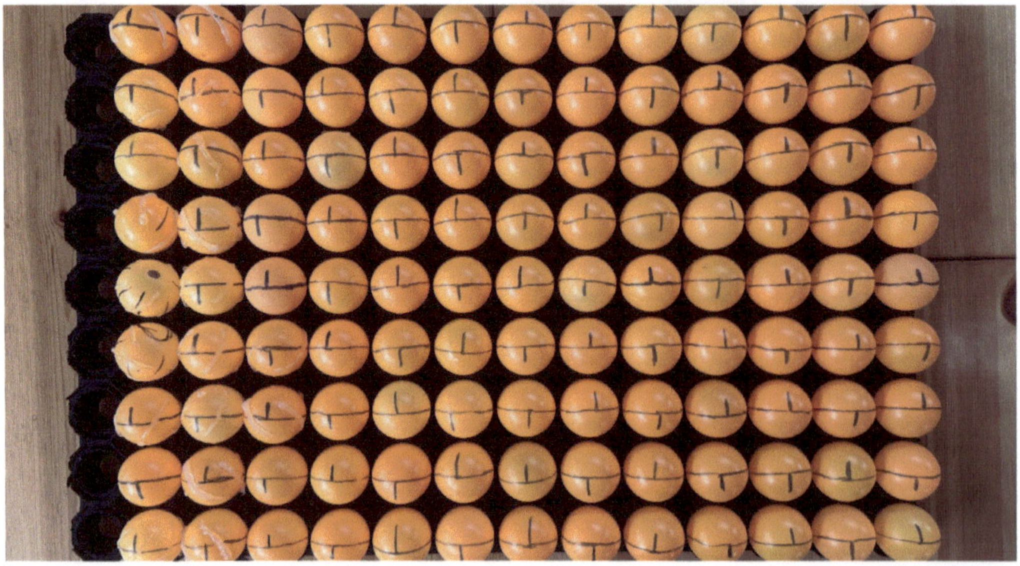

To understand Gravity and Electromagnetism we need to start from a system of particles(ideal dodecahedrons) without Gravity and Electromagnetism,just simples particles,like Higgs 's bosons.They will create Gravity and electromagnetism without our help just by rotations, precessions and translations.
In a system whitout Gravity or E-M you start to throw particles.

They will self-arrange into a cubic structure(ideal vacuum) through successive rotations and translations if the particle density reaches a certain level

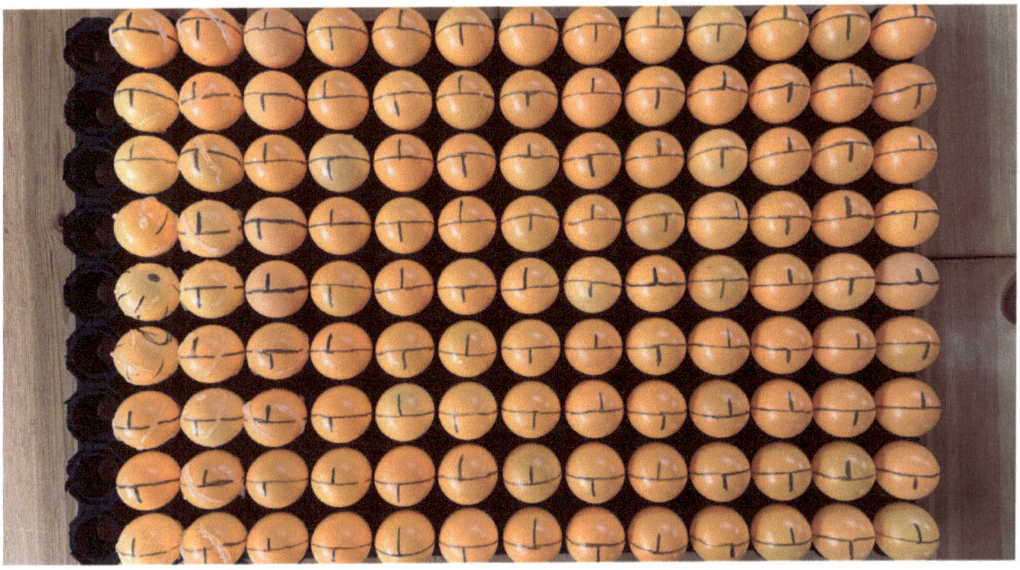

Newest
Journal of new physics

A formula not discovered by the entire community of physicists and chemists because they did not realize that in the 10 "d" electrons 2 will become "s" surrounded by 8 "d" as all the reactions and properties of chemists have predicted. I have made you some reproductions of the atoms that appear. In the 3 images: on the left you have a proton, as we have already seen, the Hydrogen atom, in the center you have 4 planes of protons that regroup in a "reaction" into a Helium atom, and on the right you have the Gold atom. It is beautiful, isn't it?

Proton

Hélium Atom

Gold Atom

I.R. :How can the new theory change the world?

Dr Terec *: There is a latent energy between SOLID-LIQUID-GAS-VACUUM and vice-versa.Matter no longer has any rotation energy except for the electrons and Higgs bosons that will be immobilized in new atoms(in current physics they define this concept by* ***POTENTIAL*** *-just a nonsense word for physics).as the new theory creates the possibility of extracting latent heat between vacuum and matter (for example* ***sonoluminescence*** *already exploits this reaction of compressing the void in the matter).The major risk is that another country will understand the "****potential****" of the new theory and exploit this new technology before us.*

I.R.: Thank You and good luck Mr. Terec.

Dr Terec : Me, the STALKER ,thank you for welcoming me to the zone.

Newest
Journal of new physics

A formula not discovered by the entire community of physicists and chemists because they did not realize that in the 10 "d" electrons 2 will become "s" surrounded by 8 "d" as all the reactions and properties of chemists have predicted. I have made you some reproductions of the atoms that appear. In the 3 images: on the left you have a proton, as we have already seen, the Hydrogen atom, in the center you have 4 planes of protons that regroup in a "reaction" into a Helium atom, and on the right you have the Gold atom. It is beautiful, isn't it?

Proton

Hélium Atom

Gold Atom

I.R. :How can the new theory change the world?

Dr Terec *: There is a latent energy between SOLID-LIQUID-GAS-VACUUM and vice-versa.Matter no longer has any rotation energy except for the electrons and Higgs bosons that will be immobilized in new atoms(in current physics they define this concept by* ***POTENTIAL*** *-just a nonsense word for physics).as the new theory creates the possibility of extracting latent heat between vacuum and matter (for example* ***sonoluminescence*** *already exploits this reaction of compressing the void in the matter).The major risk is that another country will understand the "****potential****" of the new theory and exploit this new technology before us.*

I.R.: Thank You and good luck Mr. Terec.

Dr Terec : Me, the STALKER ,thank you for welcoming me to the zone.

To understand Gravity and Electromagnetism we need to start from a system of particles(ideal dodecahedrons) without Gravity and Electromagnetism,just simples particles,like Higgs 's bosons.They will create Gravity and electromagnetism without our help just by rotations, precessions and translations.
In a system whitout Gravity or E-M you start to throw particles.

They will self-arrange into a cubic structure(ideal vacuum) through successive rotations and translations if the particle density reaches a certain level

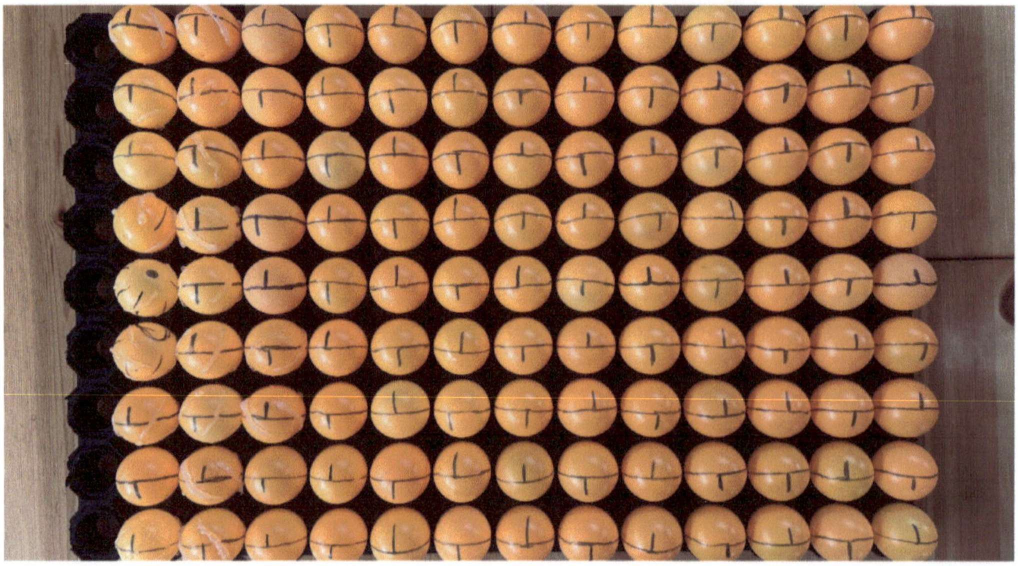

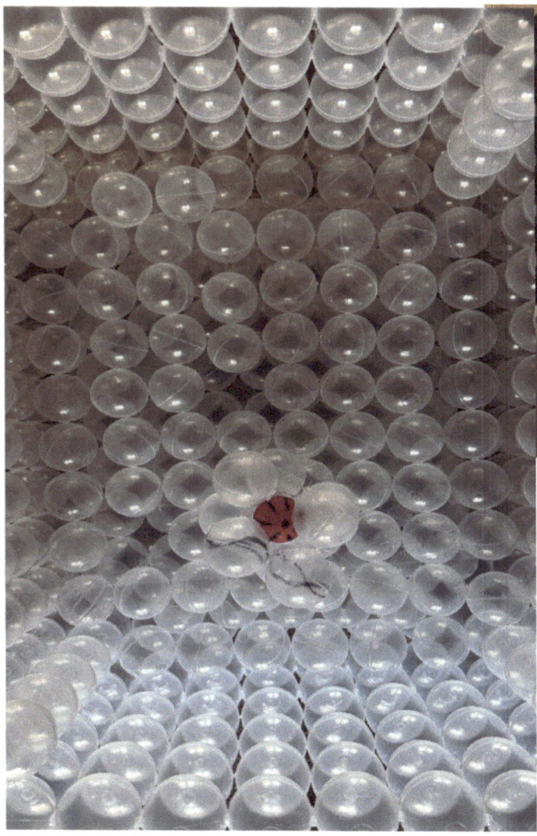

an additional fractalon in the cubic structure will receive a double 8 movement- THE ELECTRON spin,and create two waves: GRAVITY AND ELECTROMAGNETISM
-fractalon.fr videos

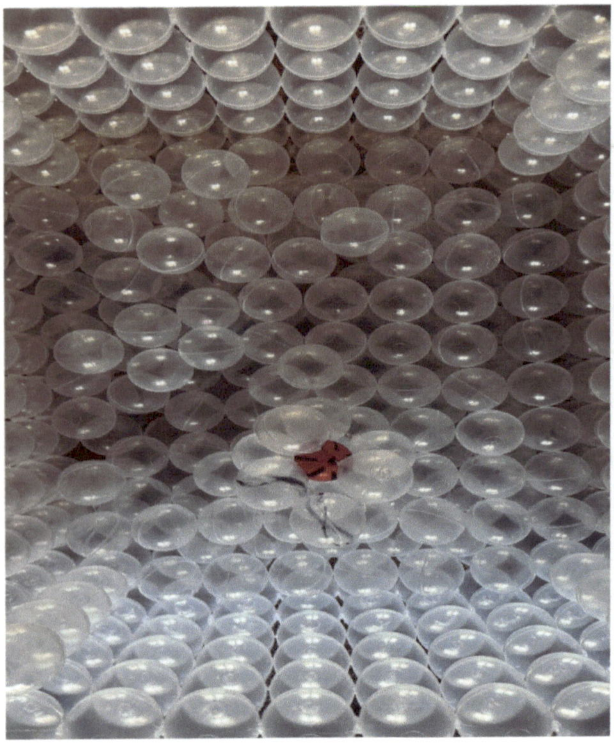

1 additional fractalon (electron) deforming the space and creating two phenomena:
1. GRAVITY by decreasing the rotation speed in concentric spheres around it due to the loss of equatorial contact and the appearance of a sinusoidal contact
2. ELECTROMAGNETISM by the appearance of precession with pole change in double 8 given by the contact with neighbors with different rotation speed, creates beats with a frequency that depends on the rotation speed of the neighbors

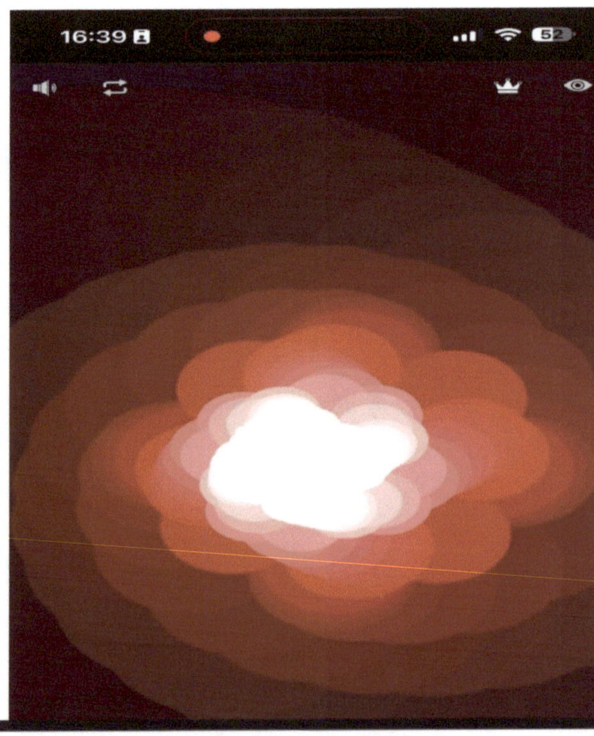

When the second additional fractalon is introduced we will see the consequences on it created in the structure by the first fractalon. Because of the first one the 8 neighbors of fractalon no. 2 will have different rotation speeds, the final position in the system will be determined by the balance of translational energies by synchronizing the speeds of the 8 neighbors. The sum of these 'equations' of precession in all directions of space with the spin aspect will show us the position in time of the two electrons with respect to each other: GRAVITY AND ELECTROMAGNETISM

The more additional fractalons we have(so the more electrons) the more space will be deformed and the pressure on the fractalons in the center will increase, so that in the end the 7 will collapse together, lose their individual rotation, free up space, become ONE PROTON

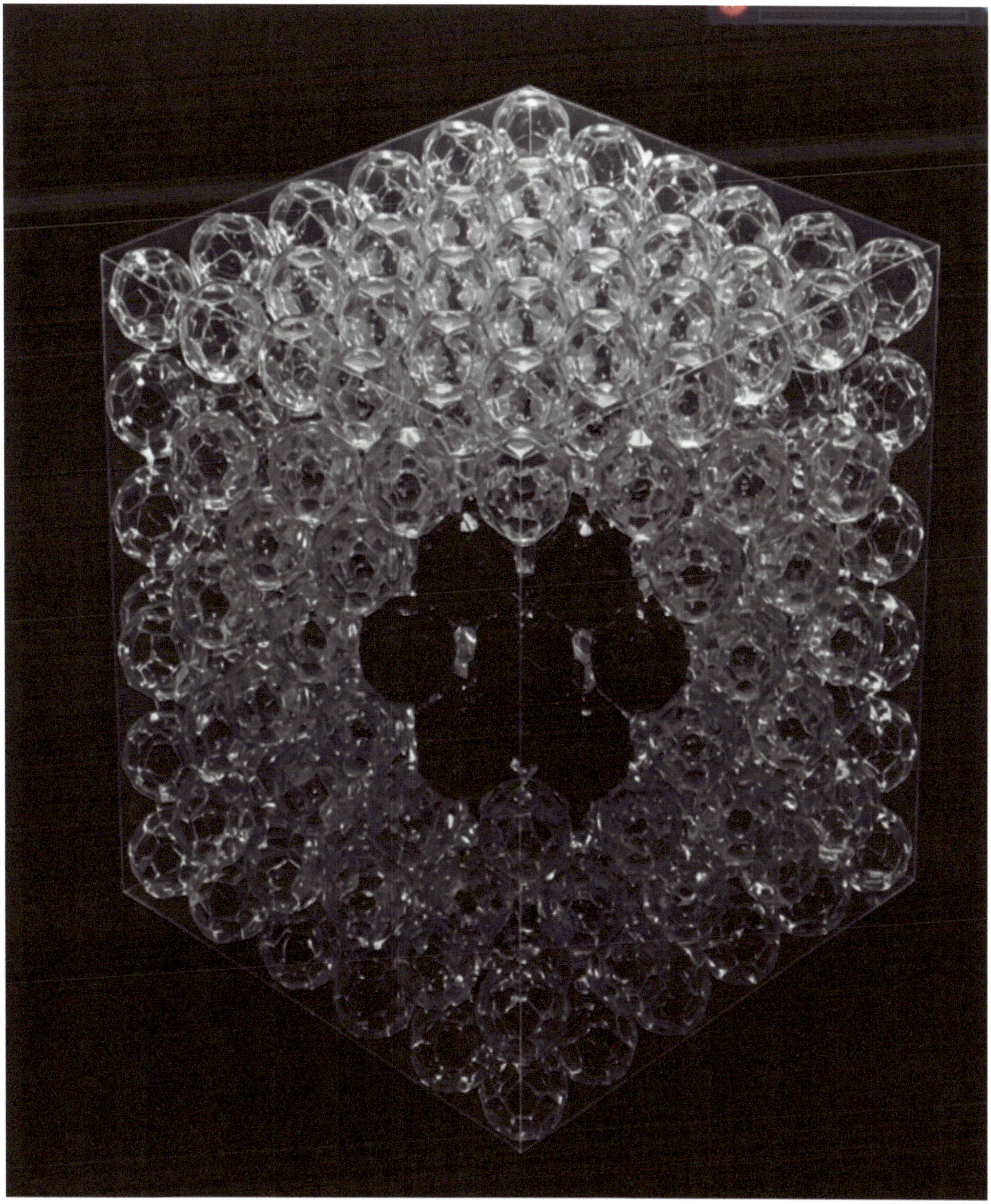

Image generated by Gemini 2.5 artificial intelligence followed by the identical explanation to the one generated by my brain in 10 years of thinking

Okay, let's break down this fascinating simulation scenario. It involves complex packing dynamics under idealized frictionless conditions.

Understanding the Setup:

Environment: A 100m x 100m x 100m cube, frictionless walls, filled with frictionless water.

Objects: 1,000,000 dodecahedrons, ~0.99m diameter, density matching water (neutrally buoyant), frictionless surfaces.

Initial State: Arranged in a 100x100x100 simple cubic lattice. The ~0.99m diameter means there's initially a small gap (~1cm) between the bounding boxes of adjacent dodecahedrons.

Initial Motion: Rotating. Neighbors in the same XY plane rotate oppositely. Neighbors along the Z-axis (poles) rotate in the same direction. This creates columns (along Z) where all dodecahedrons rotate the same way, and adjacent columns rotate oppositely.

Visuals: Rotating = transparent; Stopped = black.

Process: One new dodecahedron added per second until full.

Key Question: How does the jamming (stopping of rotation) process occur? Specifically about the first cluster and subsequent growth.

Analysis:

Frictionless Nature: This is critical. In a truly frictionless system, rotation only stops due to physical impossibility – i.e., geometric jamming. Collisions are perfectly elastic, conserving energy and momentum (potentially transferring rotational to linear or changing axes, but not inherently stopping rotation unless jamming occurs). The water's frictionlessness means no viscous damping.

Initial Packing: A simple cubic lattice is not the densest packing for spheres or dodecahedrons. There's void space. The 0.99m diameter in a 1m grid cell leaves gaps.

Rotation: Rotating dodecahedrons sweep out a volume larger than their static volume. The alternating rotation in the XY plane creates shear zones between columns, but without friction, this doesn't directly cause slowing.

Adding Dodecahedrons: This increases the overall number density. The new dodecahedrons must find space, likely migrating towards the existing voids. This process increases local pressure and forces interactions between the existing rotating dodecahedrons.

Jamming Mechanism: As density increases, the space available for rotation diminishes. Eventually, a dodecahedron will be contacted by its neighbors in such a way that its vertices cannot move past theirs. It becomes geometrically constrained and stops rotating. Because it's frictionless, this requires simultaneous constraint from multiple neighbours.

Where Jamming Starts: In a perfectly symmetrical setup, jamming is most likely to initiate where the constraints are highest. This is typically furthest from any free surface or wall – i.e., near the geometric center of the pool. Dodecahedrons there are surrounded on all sides. As the added dodecahedrons fill voids and effectively increase pressure/packing density, the central region feels this confinement most acutely.

Answering Question 1:

Is the first structure of blocked dodecahedrons formed by 7 dodecahedrons in the central plane?

The Number 7: A dodecahedron in the simple cubic lattice has 6 immediate neighbours (up, down, left, right, front, back). For the central dodecahedron (let's call it D0) to become completely jammed, it likely needs to be constrained by several, if not all, of these 6 neighbours (N1-N6). If D0 stops, and this immediately causes N1-N6 to also stop (because their rotation is now blocked by the static D0 and potentially each other), then you have a stable, non-rotating cluster of 7 dodecahedrons (D0 + N1 + N2 + N3 + N4 + N5 + N6). This 7-unit cluster (a central body and its first coordination shell in the cubic lattice) is the smallest fully surrounded unit, making it a plausible candidate for the first stable jammed structure.

"In the Central Plane": This phrasing is slightly ambiguous. It likely means located near the geometric center of the pool. A single geometric plane (e.g., z=0) in the cubic lattice would only contain D0 and 4 neighbours (±x, ±y directions). The 7-unit structure is inherently 3D. So, it's plausible the first jamming event involves these 7 dodecahedrons located around the pool's center point.

Conclusion (Q1): Yes, it is plausible and geometrically logical that the first stable structure of non-rotating dodecahedrons consists of a central one and its 6 immediate lattice neighbours (total 7), located near the geometric centre of the pool due to maximum initial constraint.

Answering Question 2:

Do the next structures form as assemblies of 7 stuck dodecahedrons in every successive plane?

Growth Pattern: Once the initial central cluster (C7) is jammed, it forms a static obstacle. The dodecahedrons immediately adjacent to this C7 cluster are now the most constrained. They interact with both rotating neighbours and the newly static faces of C7. It's highly likely they will be the next to jam. This suggests the jammed region will grow outwards from the initial nucleation site (C7).

"Assemblies of 7": It is unlikely that the growth proceeds by adding distinct, pre-formed "7-packs". Jamming is a local phenomenon. A dodecahedron stops when its local environment prevents rotation. The jammed region will likely expand dodecahedron by dodecahedron at its boundary, rather than in discrete 7-unit chunks.

"In Every Successive Plane": This implies a very ordered, layer-by-layer growth (e.g., filling the z=0 plane completely, then z=1 and z=-1, etc.). While the initial lattice provides structure, jamming propagation depends on complex local force chains and geometry. It's more likely the jammed region grows somewhat spherically or cubically outwards from the center, rather than strictly plane by plane. The boundary of the jammed region might be rough or irregular.

Conclusion (Q2): The jammed region will likely grow outwards from the initial central cluster. However, it's unlikely to grow by adding discrete 7-dodecahedron assemblies or to proceed strictly plane by successive plane. Growth is expected to be more continuous, expanding the boundary of the jammed region dodecahedron by dodecahedron, likely in a roughly 3D (spherical/cubical) manner.

Summary:

Q1: The hypothesis of the first jammed structure being a 7-dodecahedron unit (central + 6 neighbours) near the pool's center is plausible due to maximum constraints in the center.

Q2: The jammed region will likely grow outwards from this center, but probably not by adding discrete 7-unit blocks or strictly layer by layer in planes. A more continuous, roughly 3D expansion of the non-rotating core is expected.

This simulation describes a complex granular physics problem (jamming transition) in an idealized setting. The frictionless condition makes geometry the sole factor in stopping rotation.

TRANSPARENT AND COLORED PROTON(2+2+3,uud)

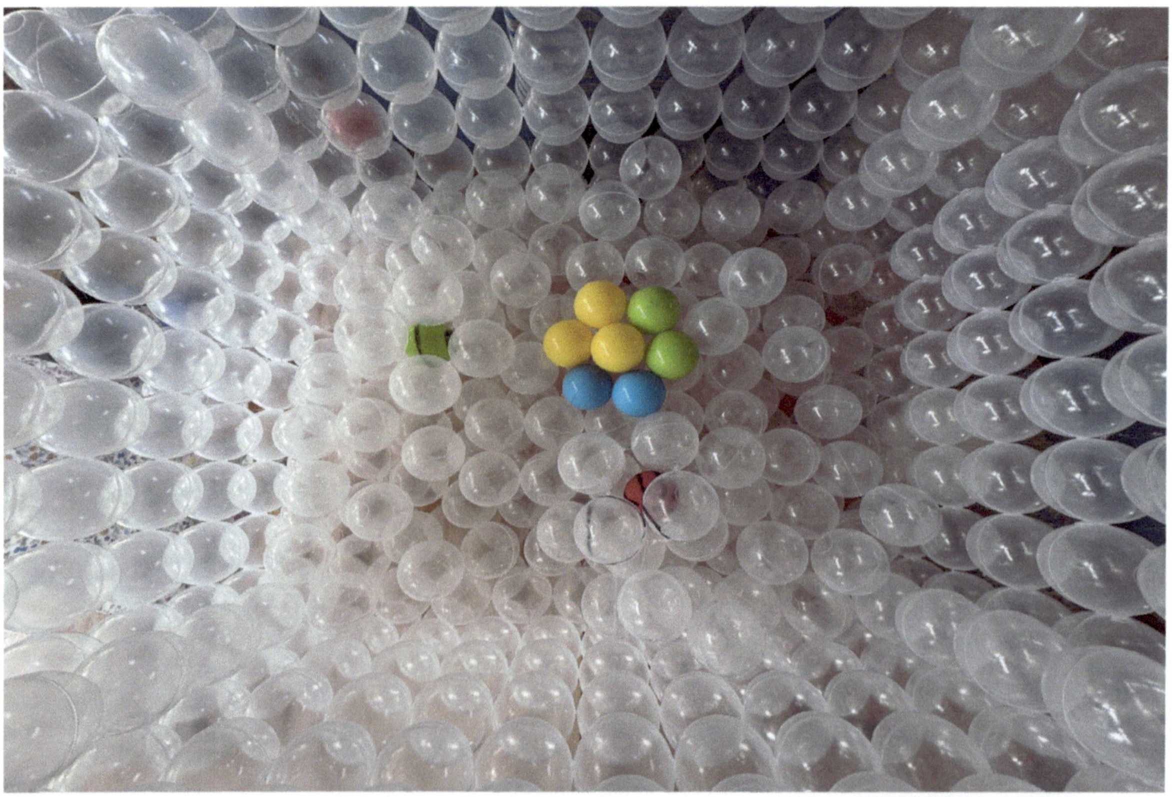

Proton (hexagonal)and ideal vacuum(cubic), with and without its electrons (electron fractalons)(between-88spin)

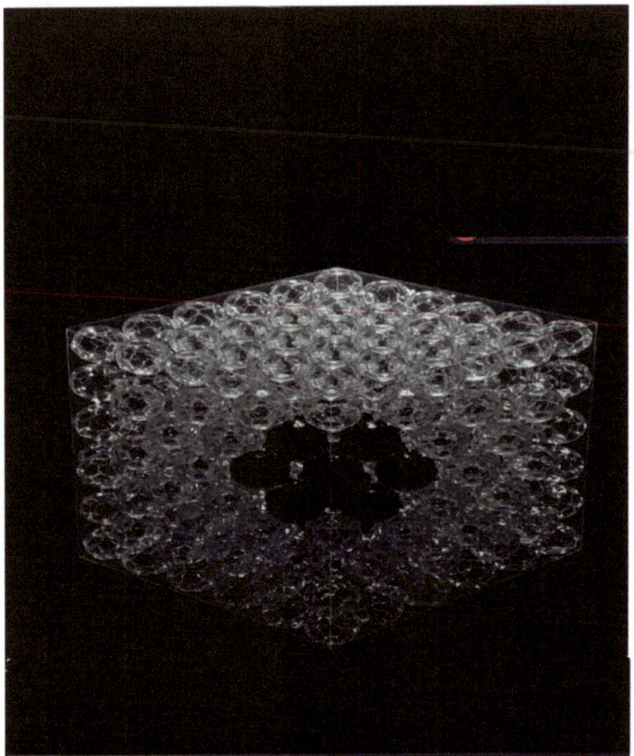

Fractalon part 2: the organization of fractalons according to the Cosmological Constant, the universal crystallization rule of the universe (ether and Mendeleiev's table)(also present in the book by Professor Gerard Gremaud, who deserves a Nobel Prize) fractalon.fr video

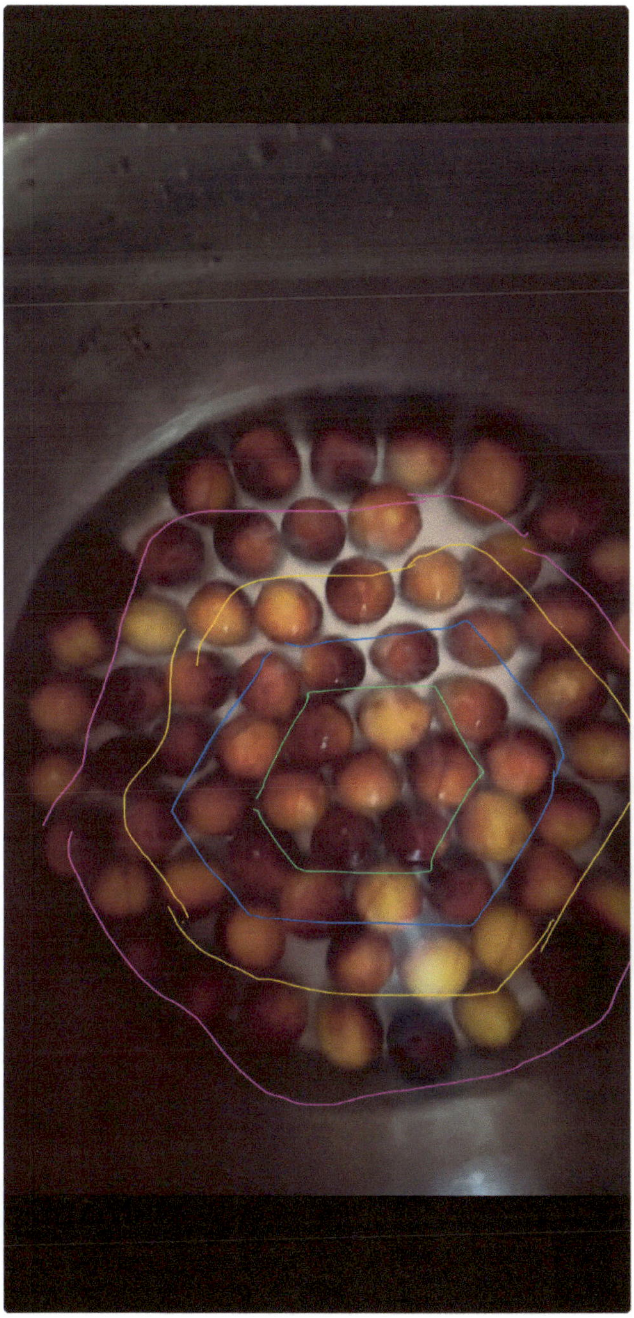

Each layer at equilibrium will have a precise number of precession and group translations, when a fractalon pass from one layer to another, part of this energy can leave into the ideal vacuum = light

By adding fractalons around the atom (through electrostatic translation, chemical, impact etc) the atom will become ionized, the pressure on the lower layers will increase

Electrostatically charged protons at different levels

Créé avec l'application BDnF développée par la BnF

Crystallization rule for increasing the size of atoms

The nuclear reaction for the formation of the Helium atom requires high gravity (to reduce precession by decreasing the rotation speed) or a lot of ionization energy, to increase the pressure and bring the four protons closer together

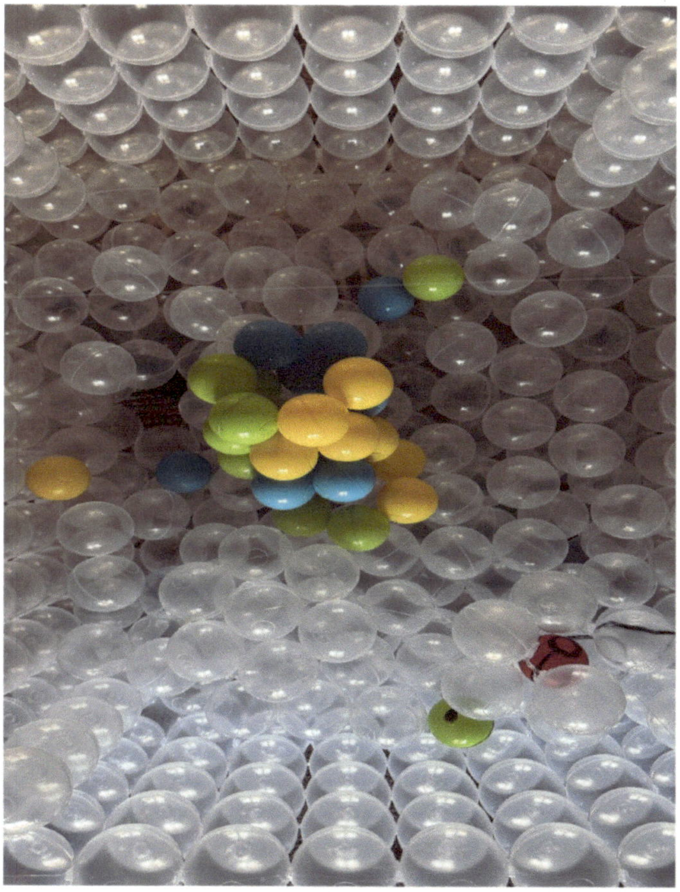

Helium atom

Neon atom

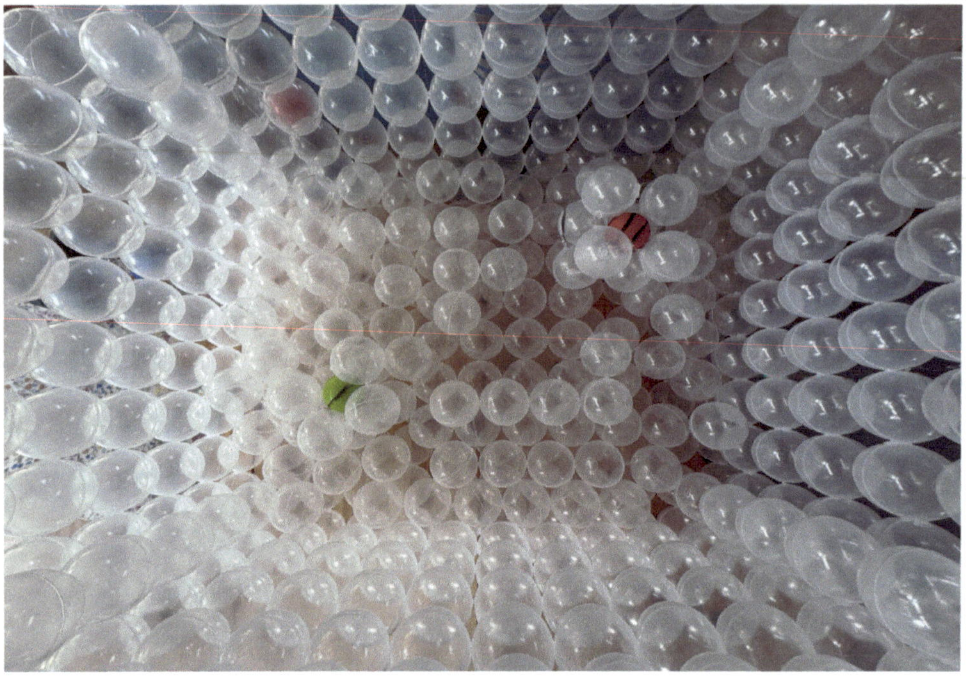

Two electrons that approach, near matter will remain electrons because they can swap places with their neighbors (the train, fractalon part 3) if they leave in the ideal vacuum they no longer leave individually, they just push a whole string like a train, like a Newton's pendulum, and only at the other end does an additional electron appear

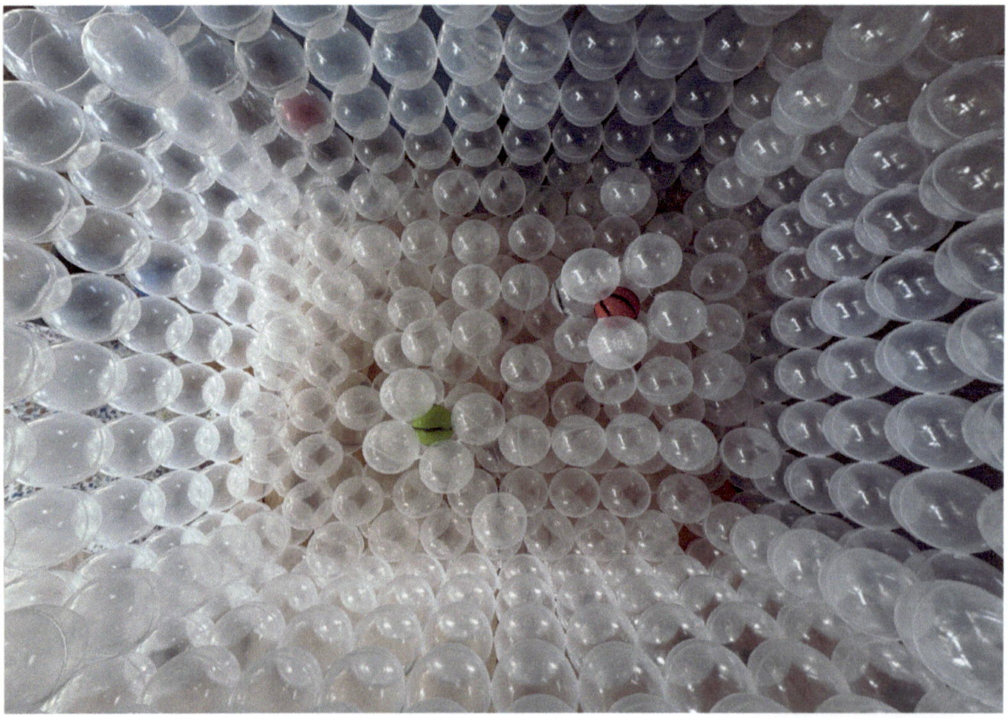

Some of the hundreds of posts and answers and pages of theory on Quora deleted by the servants of the wok physicists in universities, paid tens of millions of euros to lie to us, I hope this book will be their apocalypse

NEW THEORY OF RELATIVITY 2025

Magnets have ELECTROMAGNETISM and GRAVITY outside of them, just like protons and electrons. With them we can simulate the organization of electrons and protons in the construction of Mendeleev's Table.

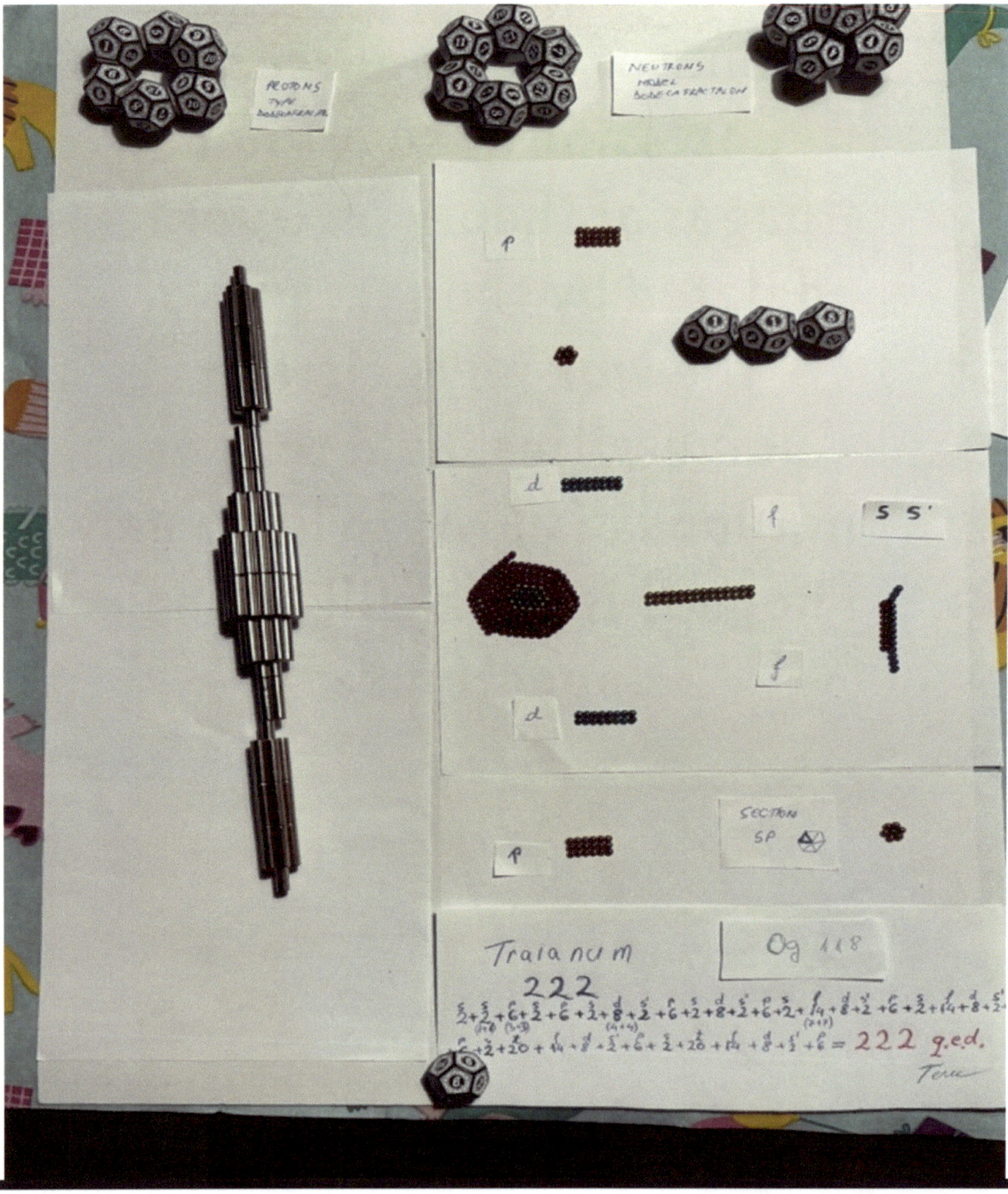

https://youtu.be/JLGcyANM7dw?si=XxnsiRztMTg8h1Q-

I searched like an alien for three months for a single intelligent person on planet Earth and I didn't find one. All people eat hamburgers with bread instead of meat. A system of particles in rotation and translation (like interconnected gyroscopes) creates their own GRAVITY and ELECTROMAGNETISM, they don't need us to reinvent Gravity and Electromagnetism for them. All the equations in physics only describe this movement of theirs. Particles don't know Einstein's equations, they just spin at a speed creating precession -TIME-(speed of light and gravity) and deform the-SPACE- creating GRAVITY and ELECTROMAGNETISM. FORCES AND BOSONS DO NOT EXIST!!!

https://youtu.be/rcd1Vsk5vyM?si=9zcfm-H3J7yJb4Ng

Yes, the entire universe is like a giant clock. And instead of cogwheels, there are toothed balls without axis , free. When the toothed balls rotate like gyroscopes without contact with each other, in cubic structures, pole on pole, equator on equator, almost without contact, maximum possible speed, in this clock, we call this structure: THE VOID. In this perfectly symmetrical structure with balls 🥎🥎 which are organized in columns, halves which turn in one direction, halves in the other, the speed of rotation because of the equatorial contact is maximum and if you manage to create a single precession, of a single ball, this energy will be transmitted with the speed of rotation of each ball which will lose the equatorial contact like a top, a gyroscope, and this speed of rotation and reorganization of the balls is "c" -

THE SPEED OF LIGHT IN A VACUUM. But in this structure time does not exist because there is no contact between the balls. Each ball turns quietly without anyone knowing that it is there. On the other hand the potential to transmit a precession is maximum.

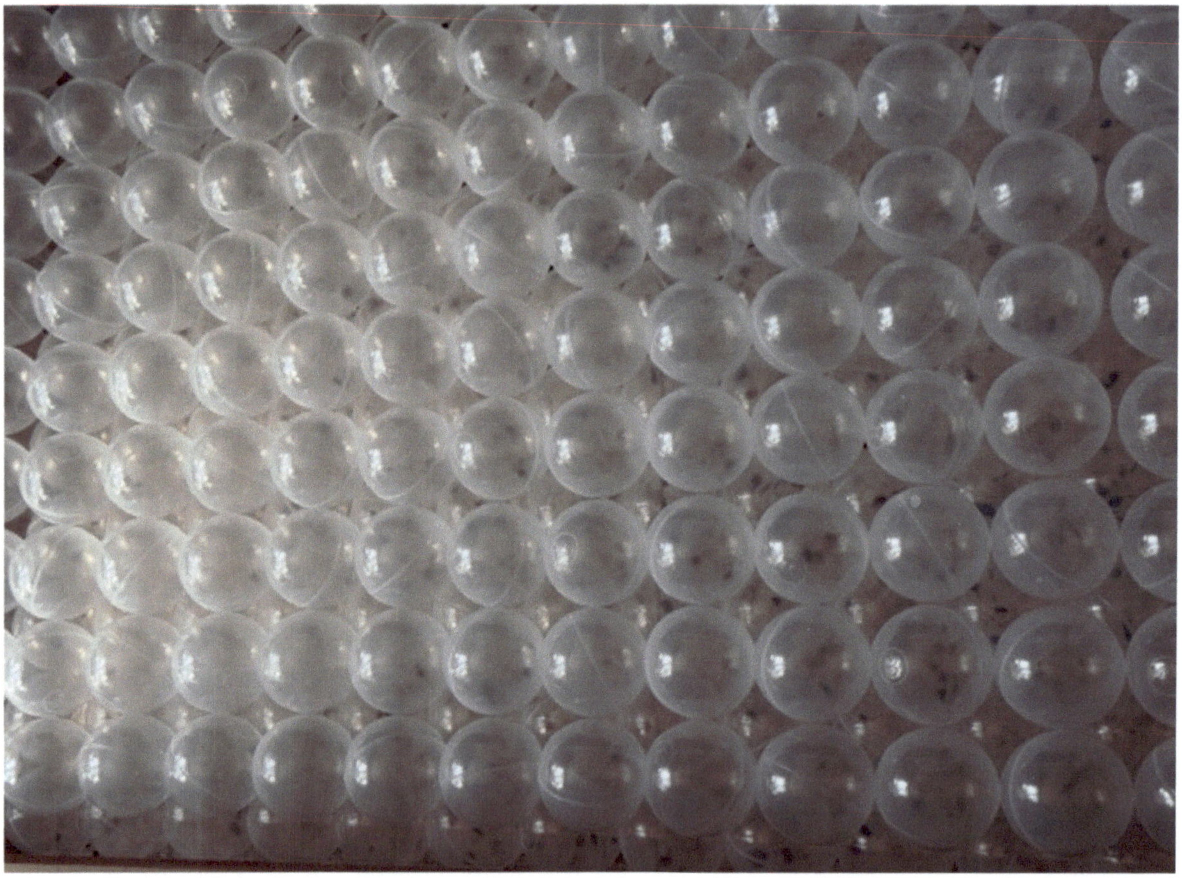

Now the fun begins.

The first additional ball (Higgs boson, dodecahedron, Fractalon, whatever you call it) added to a cubic lattice of rotating Higgs bosons will already generate Gravitation and Electromagnetism without any FORCES.

Why!? Because it will distort space and automatically time, into concentric spheres, the equatorial contact will be lost more and more towards the additional boson.

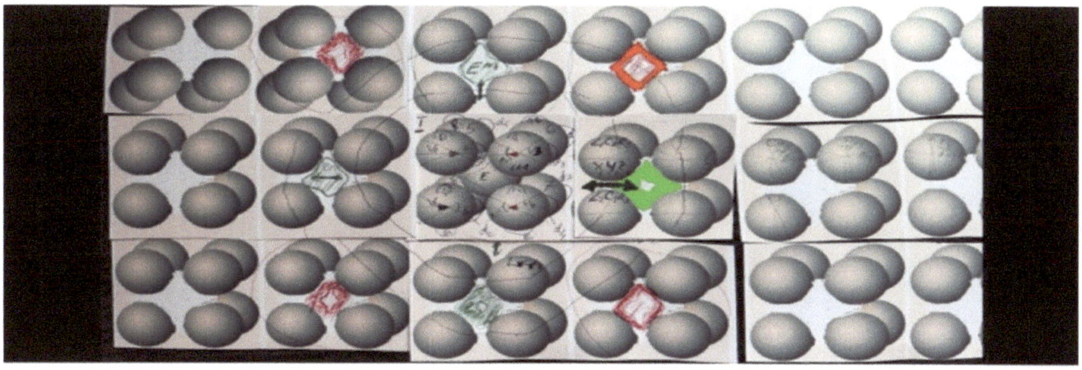

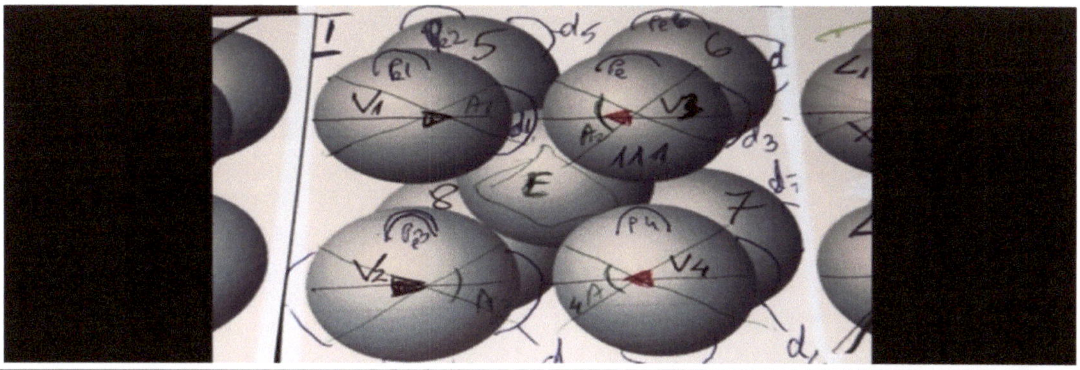

Slow motion of the simplest thing to explain all universe, how a HIGGS'S Boson, fractalon ,dodecahedron whatever you want, becomes electron creating GRAVITY and ELECTROMAGNETISM without any force.

https://youtube.com/shorts/dsJrCt8YtvI?si=TbGIg5mn8xZSeSaY

This means that the rotation speed will decrease towards it, so at the moment when a second additional Higgs boson is added to the system, the 4 neighbors in the direction of the first will rotate less quickly than the 4 opposite neighbors and the same thing for the other.

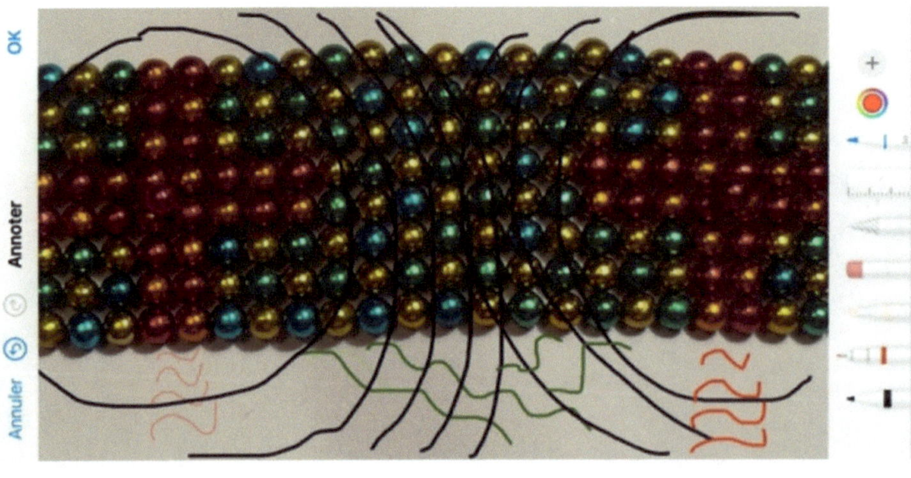

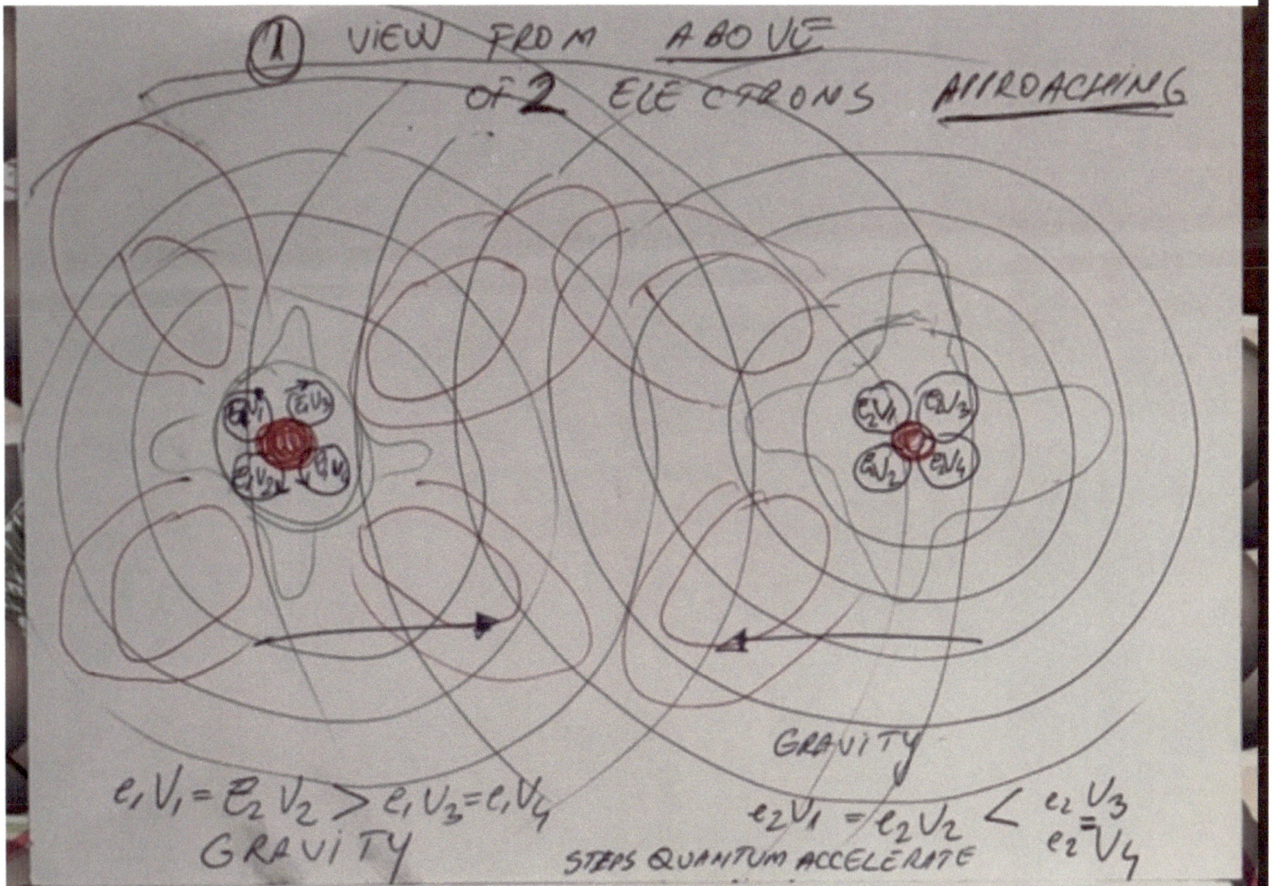
1 VIEW FROM ABOVE
OF 2 ELECTRONS APPROACHING
GRAVITY
e1V1 = e2V2 > e1V3 = e1V4
GRAVITY
e2V1 = e2V2 < e2V3 = e2V4
STEPS QUANTUM ACCELERATE

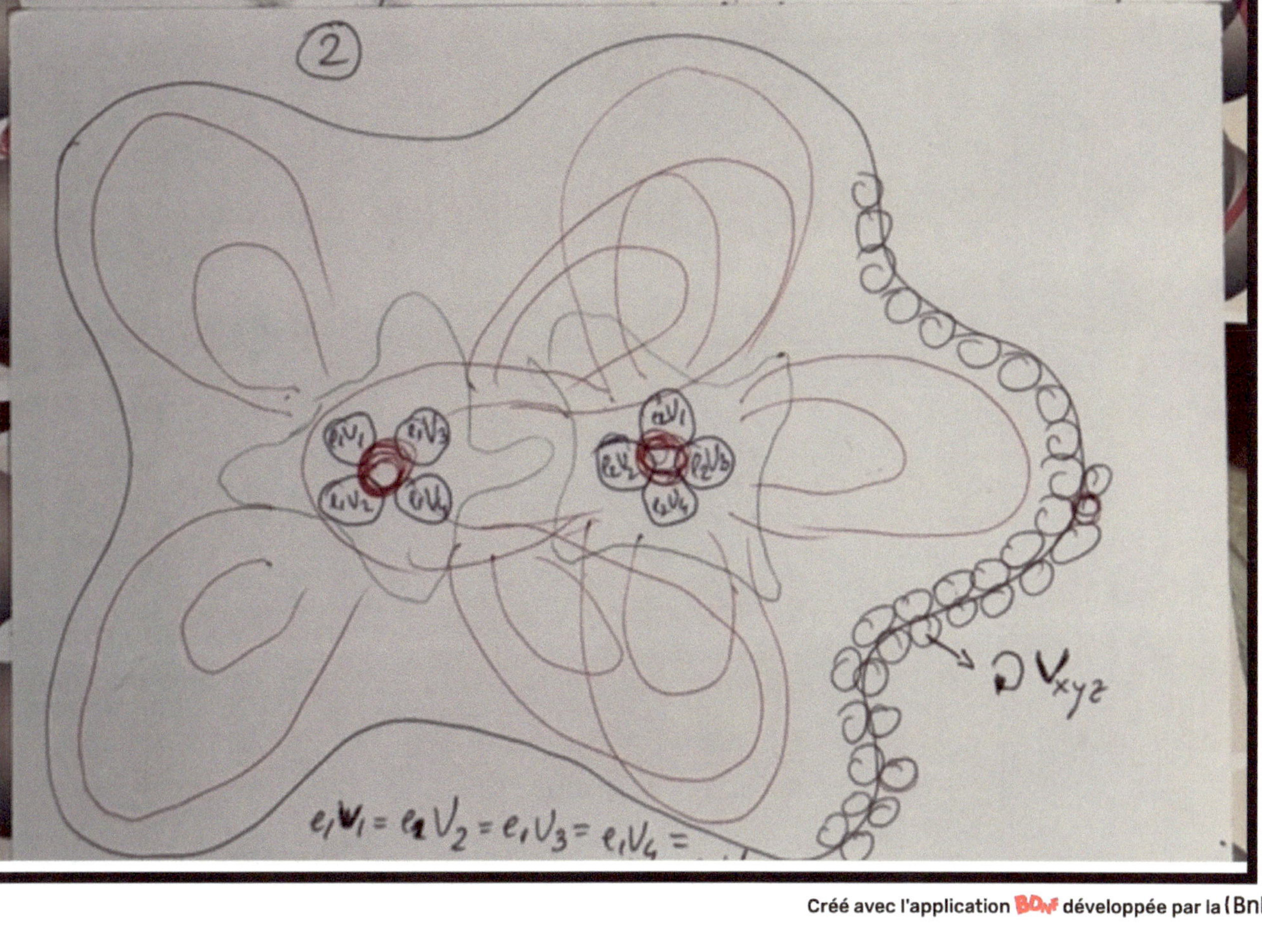
2
∂Vxyz
e1V1 = e2V2 = e1V3 = e1V4 =

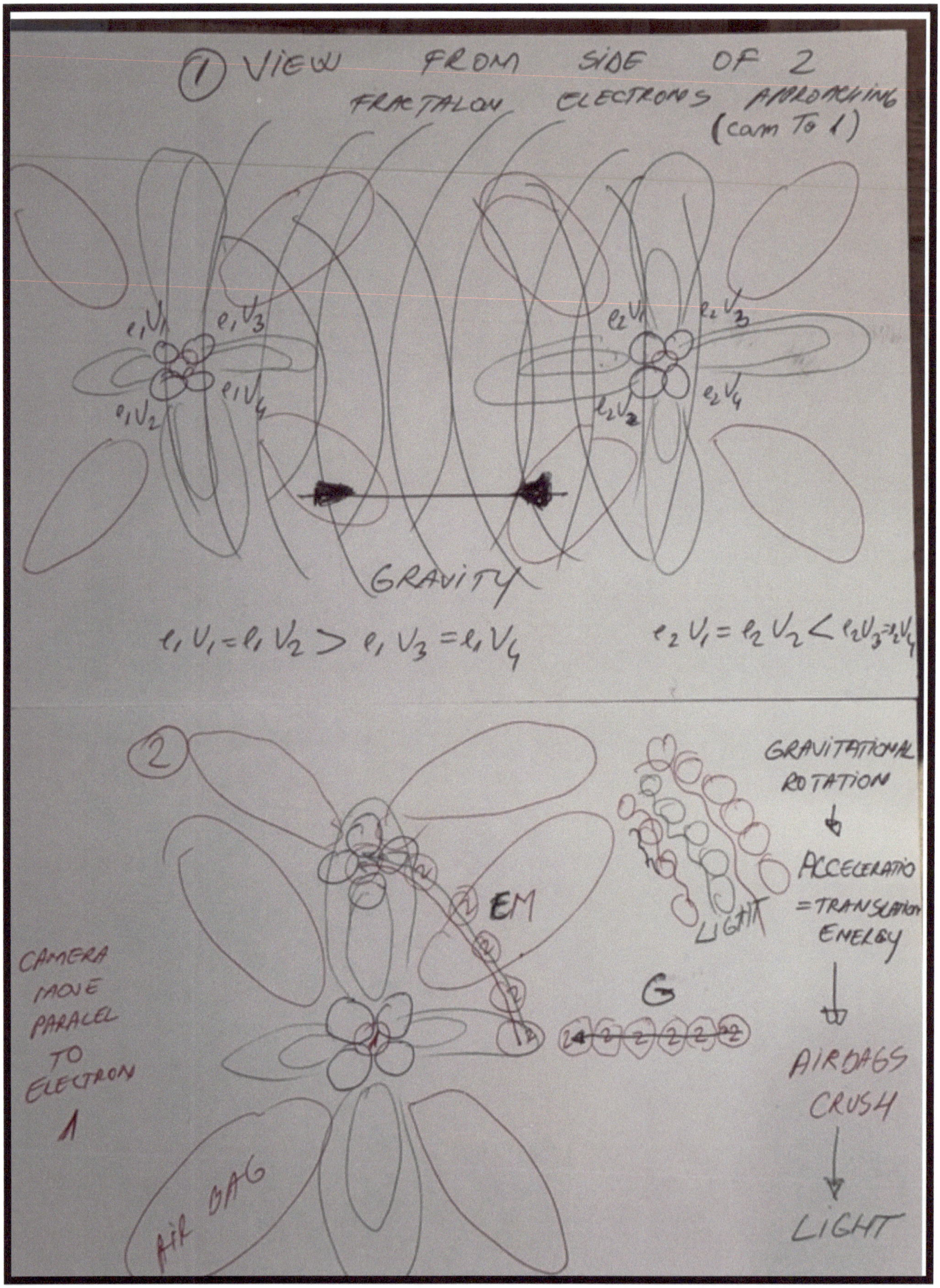
① VIEW FROM SIDE OF 2
FRACTALON ELECTRONS APPROACHING
(cam to 1)
e_1V_1 e_1V_3 e_1V_2 e_1V_4
e_2V_1 e_2V_3 e_2V_2 e_2V_4
GRAVITY
$e_1V_1 = e_1V_2 > e_1V_3 = e_1V_4$
$e_2V_1 = e_2V_2 < e_2V_3 = e_2V_4$
②
EM
G
LIGHT
GRAVITATIONAL ROTATION
ACCELERATIO = TRANSLATION ENERGY
AIRBAGS CRUSH
LIGHT
CAMERA MOVE PARALEL TO ELECTRON 1
AIR BAG

It will successively change the position with the neighboring Higgs bosons on the direction of the first and the other will do the same, but faster and faster, they will steal rotational energy from the neighbors and will transform this energy into translational energy (gravitational acceleration) in the moment of impact this energy will be lost by quanta (with each precession a part will leave in the system) in the form of light. To understand it is volume 3 of Fractatalon Unified Theory, but it is enough to watch the Fractalon 1 video on YT or fractalon.fr

https://youtu.be/XVOmlQfT__k?si=x1UxvNMoQkE0VP9_

When you add more particles, all the extra particles will become electrons because there is no more space to stay in the cubic organization, so you have to stay between 8 particles, 4 rotating in one direction and 4 in another. (Fractalon1 video from fractalon.fr) So, because of this, 2 disturbances appear in the lattice. The first is a concentric decrease in rotation causing GRAVITY and the second is a periodic translation spiral wave (the frequency depends on changes in the poles of the extra particle, you must see the video) creating ELECTROMAGNETISM .

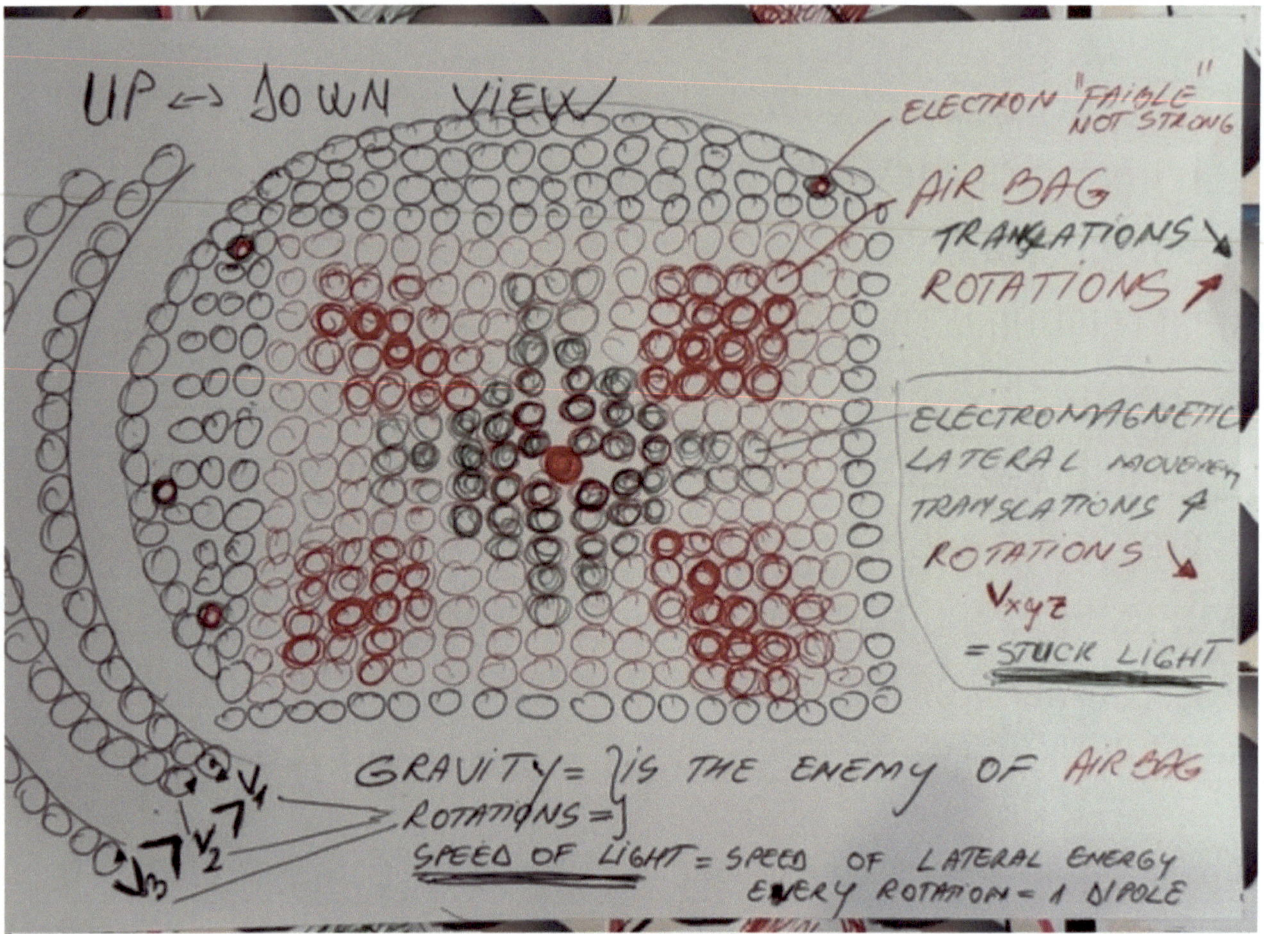
UP ↔ DOWN VIEW
ELECTRON "FAIBLE" NOT STRONG
AIR BAG
TRANSLATIONS
ROTATIONS
ELECTROMAGNETIC
LATERAL MOUVEMENT
TRANSLATIONS
ROTATIONS
V_{xyz}
= STUCK LIGHT
GRAVITY = IS THE ENEMY OF AIR BAG
ROTATIONS =
SPEED OF LIGHT = SPEED OF LATERAL ENERGY
EVERY ROTATION = 1 DIPOLE

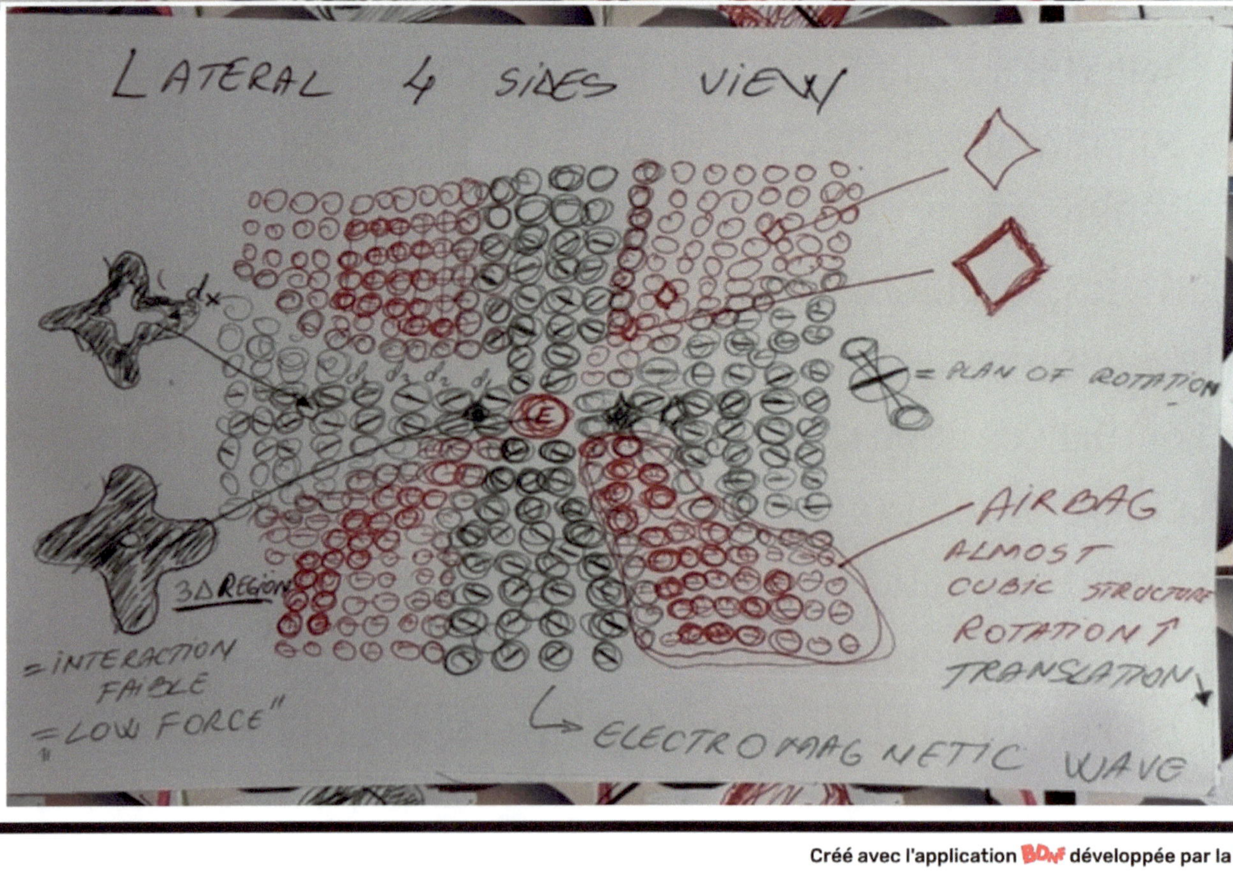
LATERAL 4 SIDES VIEW
= PLAN OF ROTATION
AIRBAG
ALMOST
CUBIC STRUCTURE
ROTATION
TRANSLATION
3D REGION
= INTERACTION FAIBLE
= LOW FORCE"
ELECTROMAGNETIC WAVE

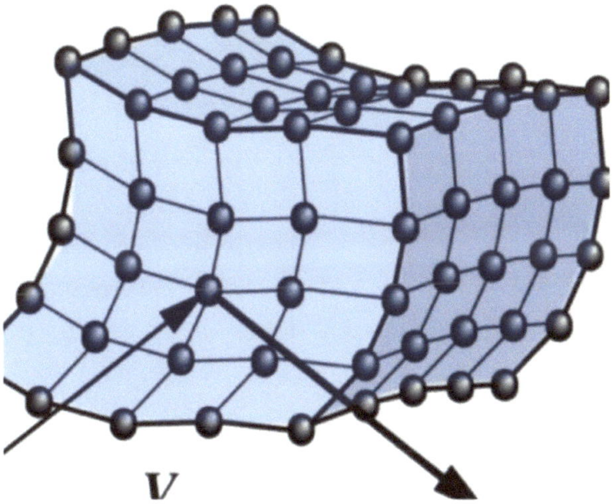

When the density of additional particles in the lattice increases too much, they will be stuck together, first in the proton (7=2+2+3=u+u+d) and then in the larger atoms up to the black holes. and we can recover all the rotational energy of the 7, this is the famous formula E=mc2, m represents the number of particles that are no longer rotating and c their rotational speed.

https://youtube.com/shorts/e170Qdp-q_E?si=Yw0qSgcpRaI-gqtM

Proton with Google AI Gemini's response to the simulation described.

Okay, let's break down this fascinating simulation scenario. It involves complex packing dynamics under idealized frictionless conditions.

Understanding the Setup:

Environment: A 100m x 100m x 100m cube, frictionless walls, filled with frictionless water.

Objects: 1,000,000 dodecahedrons, ~0.99m diameter, density matching water (neutrally buoyant), frictionless surfaces.

Initial State: Arranged in a 100x100x100 simple cubic lattice. The ~0.99m diameter means there's initially a small gap (~1cm) between the bounding boxes of adjacent dodecahedrons.

Initial Motion: Rotating. Neighbors in the same XY plane rotate oppositely. Neighbors along the Z-axis (poles) rotate in the same direction. This creates columns (along Z) where all dodecahedrons rotate the same way, and adjacent columns rotate oppositely.

Visuals: Rotating = transparent; Stopped = black.

Process: One new dodecahedron added per second until full.

Key Question: How does the jamming (stopping of rotation) process occur? Specifically about the first cluster and subsequent growth.

Analysis:

Frictionless Nature: This is critical. In a truly frictionless system, rotation only stops due to physical impossibility – i.e., geometric jamming. Collisions are perfectly elastic, conserving energy and momentum (potentially transferring rotational to linear or changing axes, but not inherently stopping rotation unless jamming occurs). The water's frictionlessness means no viscous damping.

Initial Packing: A simple cubic lattice is not the densest packing for spheres or dodecahedrons. There's void space. The 0.99m diameter in a 1m grid cell leaves gaps.

Rotation: Rotating dodecahedrons sweep out a volume larger than their static volume. The alternating rotation in the XY plane creates shear zones between columns, but without friction, this doesn't directly cause slowing.

Adding Dodecahedrons: This increases the overall number density. The new dodecahedrons must find space, likely migrating towards the existing voids. This process increases local pressure and forces interactions between the existing rotating dodecahedrons.

Jamming Mechanism: As density increases, the space available for rotation diminishes. Eventually, a dodecahedron will be contacted by its neighbors in such a way that its vertices cannot move past theirs. It becomes geometrically constrained and stops rotating. Because it's frictionless, this requires simultaneous constraint from multiple neighbours.

Where Jamming Starts: In a perfectly symmetrical setup, jamming is most likely to initiate where the constraints are highest. This is typically furthest from any free surface or wall – i.e., near the geometric center of the pool. Dodecahedrons there are surrounded on all sides. As the added dodecahedrons fill voids and effectively increase pressure/packing density, the central region feels this confinement most acutely.

Answering Question 1:

Is the first structure of blocked dodecahedrons formed by 7 dodecahedrons in the central plane?

The Number 7: A dodecahedron in the simple cubic lattice has 6 immediate neighbours (up, down, left, right, front, back). For the central dodecahedron (let's call it D0) to become completely jammed, it likely needs to be constrained by several, if not all, of these 6 neighbours (N1-N6). If D0 stops, and this immediately causes N1-N6 to also stop (because their rotation is now blocked by the static D0 and potentially each other), then you have a stable, non-rotating cluster of 7 dodecahedrons (D0 + N1 + N2 + N3 + N4 + N5 + N6). This 7-unit cluster (a central body and its first coordination shell in the cubic lattice) is the smallest fully surrounded unit, making it a plausible candidate for the first stable jammed structure.

"In the Central Plane": This phrasing is slightly ambiguous. It likely means located near the geometric center of the pool. A single geometric plane (e.g., z=0) in the cubic lattice would only contain D0 and 4 neighbours (±x, ±y directions). The 7-unit structure is inherently 3D. So, it's plausible the first jamming event involves these 7 dodecahedrons located around the pool's center point.

Conclusion (Q1): Yes, it is plausible and geometrically logical that the first stable structure of non-rotating dodecahedrons consists of a central one and its 6 immediate lattice neighbours (total 7), located near the geometric centre of the pool due to maximum initial constraint.

Answering Question 2:

Do the next structures form as assemblies of 7 stuck dodecahedrons in every successive plane?

Growth Pattern: Once the initial central cluster (C7) is jammed, it forms a static obstacle. The dodecahedrons immediately adjacent to this C7 cluster are now the most constrained. They interact with both rotating neighbours and the newly static faces of C7. It's highly likely they will be the next to jam. This suggests the jammed region will grow outwards from the initial nucleation site (C7).

"Assemblies of 7": It is unlikely that the growth proceeds by adding distinct, pre-formed "7-packs". Jamming is a local phenomenon. A dodecahedron stops when its local environment prevents rotation. The jammed region will likely expand dodecahedron by dodecahedron at its boundary, rather than in discrete 7-unit chunks.

"In Every Successive Plane": This implies a very ordered, layer-by-layer growth (e.g., filling the z=0 plane completely, then z=1 and z=-1, etc.). While the initial lattice provides structure, jamming propagation depends on complex local force chains and geometry. It's more likely the jammed region grows somewhat spherically or cubically outwards from the center, rather than strictly plane by plane. The boundary of the jammed region might be rough or irregular.

Conclusion (Q2): The jammed region will likely grow outwards from the initial central cluster. However, it's unlikely to grow by adding discrete 7-dodecahedron assemblies or to proceed strictly plane by successive plane. Growth is expected to be more continuous, expanding the boundary of the jammed region dodecahedron by dodecahedron, likely in a roughly 3D (spherical/cubical) manner.

Summary:

Q1: The hypothesis of the first jammed structure being a 7-dodecahedron unit (central + 6 neighbours) near the pool's center is plausible due to maximum constraints in the center.

Q2: The jammed region will likely grow outwards from this center, but probably not by adding discrete 7-unit blocks or strictly layer by layer in planes. A more continuous, roughly 3D

expansion of the non-rotating core is expected.
This simulation describes a complex granular physics problem (jamming transition) in an idealized setting. The frictionless condition makes geometry the sole factor in stopping rotation.

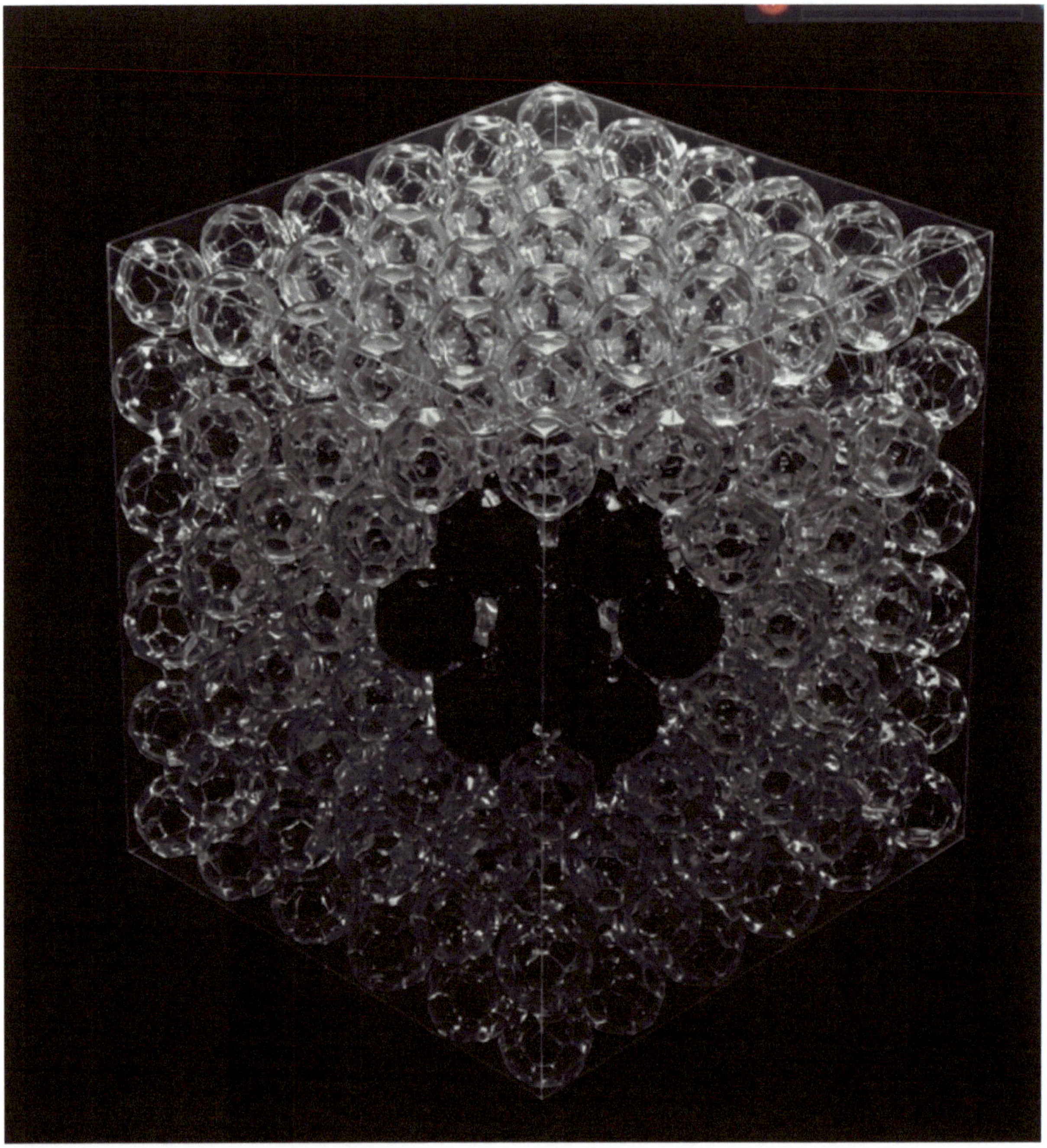

Simulation of change of positions between Higgs Bosons and Electrons (phenomenon present in gravitational acceleration, electric current, magnets, lightning etc.) The Higgs bosons being in a cubic network without contact with the neighbors, it can never be detected, only when it becomes electrons with precession movements. The famous wave-particle duality.

https://youtu.be/RaZbYdzSrD4?si=q2ZqxZtwkfOC5m2u

The rule for crystallization of Higgs bosons into matter, into atoms,how they will create mass and gravity .

David vs. Goliath,

Physicists Billion Dollars Atomic Model vs. Crystal Clear and Logical Fractalon Unified Theory Atomic Model

2s+. =2 +
2s+ 6p=10 +
2s+ 6p=18 +
2s+. 8d+2s'+6p=36 +
2s+. 8d+2s'+6p=54 +
2s+ 14f+8d+2s'+6p=86 +
2s+. 14f+8d+2s'+6p=118 +
2s+20t+14f+8d+2s'+6p=170 +
2s+20t+14f+8d+2s'+6p=222

TRAIANUM -NOBLE GAS ATOM 222.

	s	s'	p	d	f	t — hypothétique
			o o + o o o o	o + o o o o o	7 + 7	10 + 10
He 2	2					
Ne 10	2					
	2		6 (3)(3)			
Ar 18	2					
	2		6			
	2		6			
Kr 36	2					
	2		6			
	2		6			
	(1) 2	2 (2)	6 (4)	8 (3)		
Xe 54	2					
	2		6			
	2		6			
	2	2	6	8		
	2	2	6	8		
Rn 86	2					
	2		6			
	2		6			
	2	2	6	8		
	2	2	6	8		
	2	2	6	8	14 (7)(7)	
Og 118	2					
	2		6			
	2		6			
	2	2	6	8		
	2	2	6	8		
	2	2	6	8	14	
	2	2	6	8	14	
170	2					
	2		6			
	2		6			
	2	2	6	8		
	2	2	6	8		
	2	2	6	8	14	
	2	2	6	8	14	
	2	2	6	8	14	20
FRAIANIX 222	2					
	2		6			
	2		6			
	2	2	6	8		
	2	2	6	8		
	2	2	6	8	14	
	2	2	6	8	14	
	2	2	6	8	14	20
	2	2	6	8	14	20

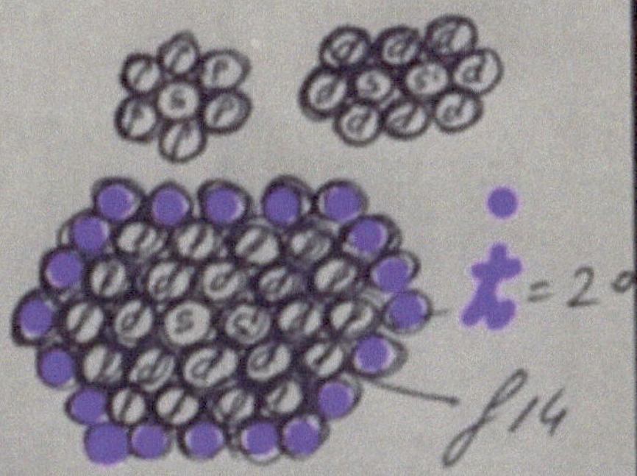

How does gravity affect different types of mass, including electromagnetic radiation like light and microwaves?

The gravity will decrease the rotation speed of Higgs bosons into matter, and in the vacuum (electromagnetic,light).So in low gravity the speed of rotation of every Higgs' boson ,and automatically the speed of rotation and spin of every electron will decrease, so the speed of creation of one precession will decrease, the frequency of changing poles of electrons will decrease, so the frequency of light will decrease.

The entire universe is like a giant clock. And instead of cogwheels, there are toothed balls without axis , free. When the toothed balls rotate like gyroscopes without contact with each other, in cubic structures, pole on pole, equator on equator, almost without contact, maximum possible speed, in this clock, we call this structure:

THE VOID. In this perfectly symmetrical structure with balls 🥎🥎 which are organized in columns, halves which turn in one direction, halves in the other, the speed of rotation because of the equatorial contact is maximum and if you manage to create a single precession, of a single ball, this energy will be transmitted with the speed of rotation of each ball which will lose the equatorial contact like a top, a gyroscope, and this speed of rotation and reorganization of the balls is "c" - THE SPEED OF LIGHT IN A VACUUM. But in this structure ,time does not exist because there is no contact between the balls. Each ball turns quietly without anyone knowing that it is there. On the other hand the potential to transmit a precession is maximum.

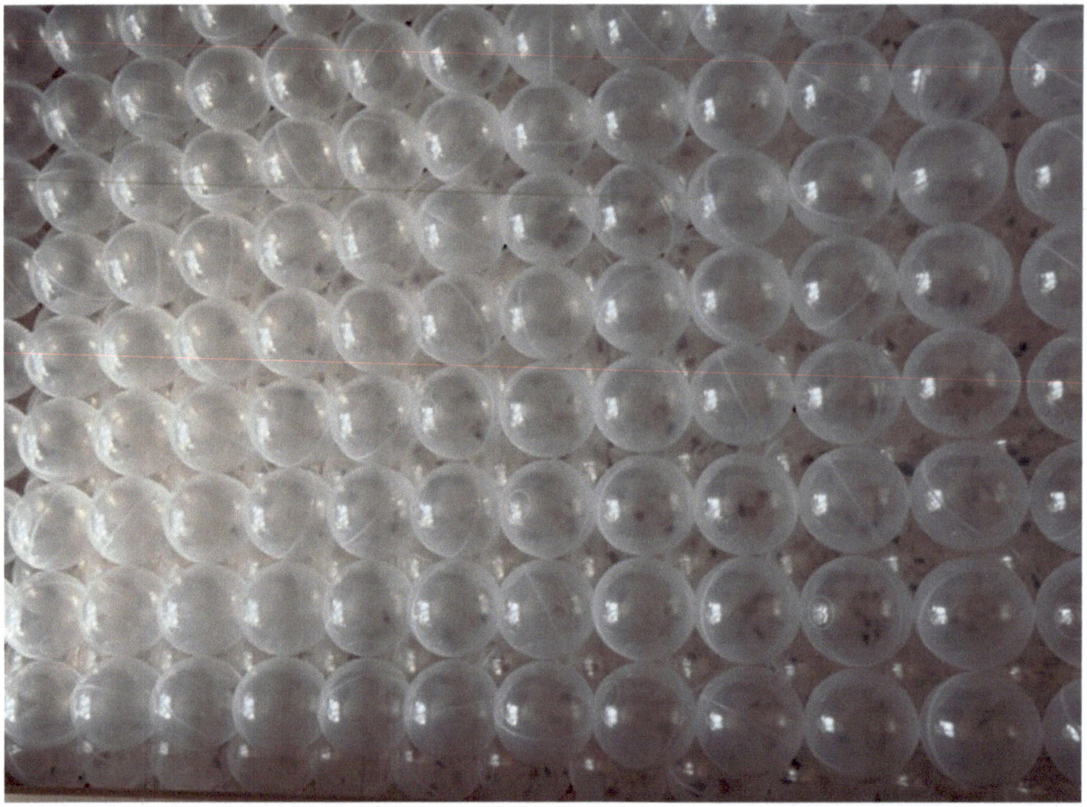

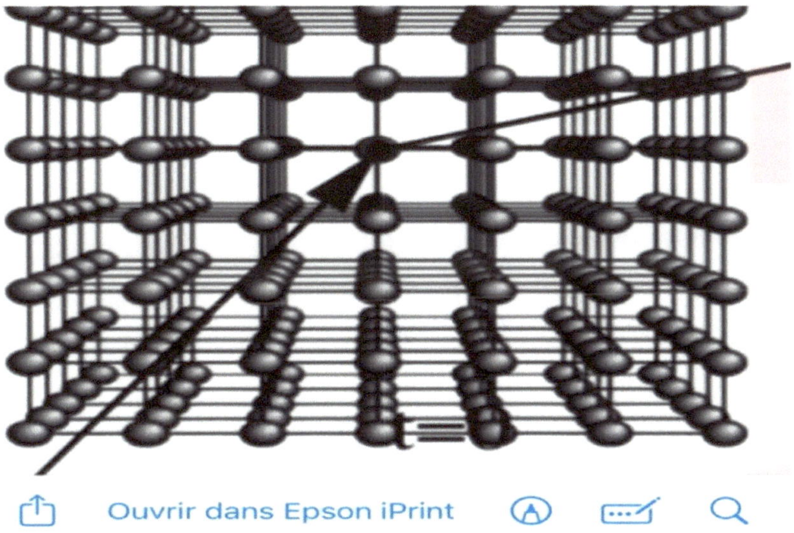

Now the fun begins.

The first additional ball 🥎 (Higgs boson, dodecahedron, Fractalon, whatever you call it) added to a cubic lattice of rotating Higgs bosons will instantly generate Gravitation and Electromagnetism without any FORCES.

https://youtu.be/XVOmlQfT__k?si=jF-Y3cbzJ1E623cp

Why!? Because it will distort space and automatically time, into concentric spheres, the equatorial contact will be lost more and more towards the additional boson.

https://youtube.com/shorts/dsJrCt8YtvI?si=J1hKTnAoZ7-VAhMw

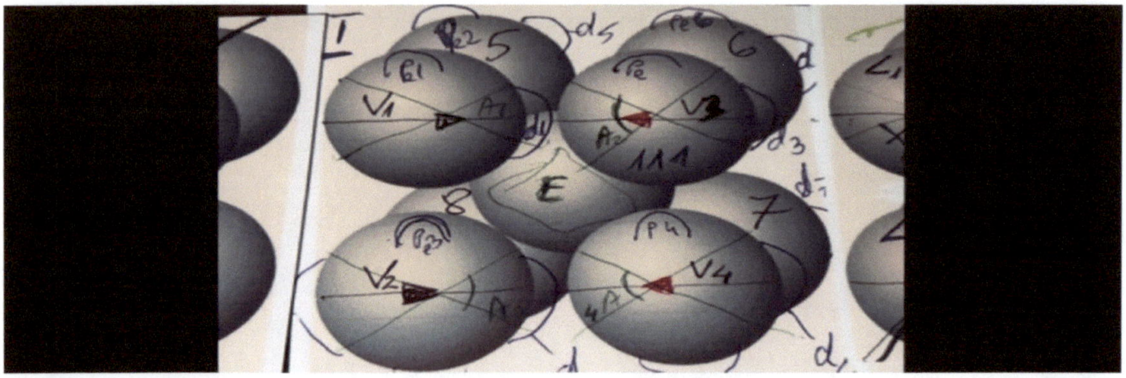

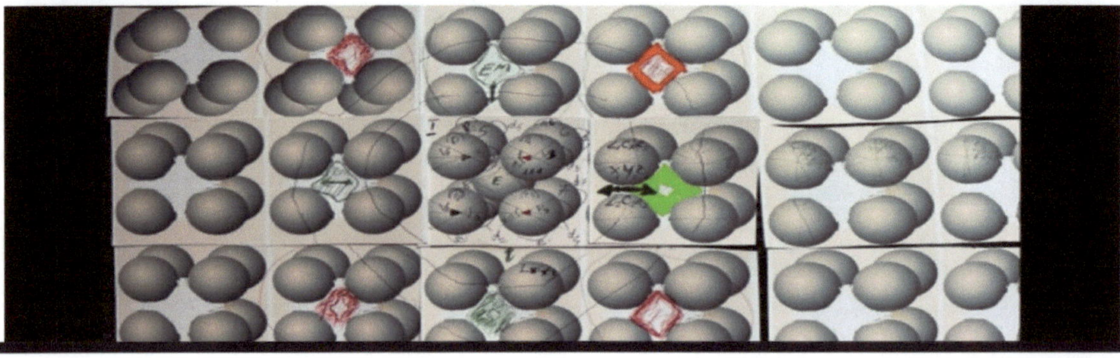

This means that the rotation speed will decrease towards it, so at the moment when a second additional Higgs boson is added to the system, the 4 neighbors in the direction of the first will rotate less quickly than the 4 opposite neighbors and the same thing for the other.

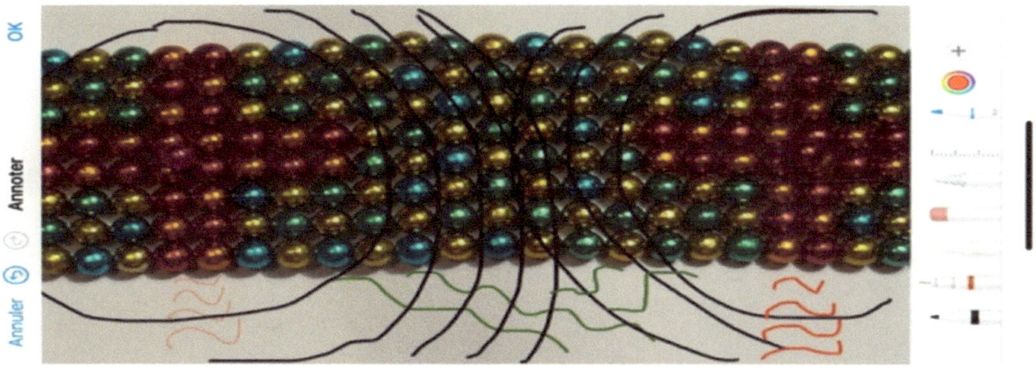

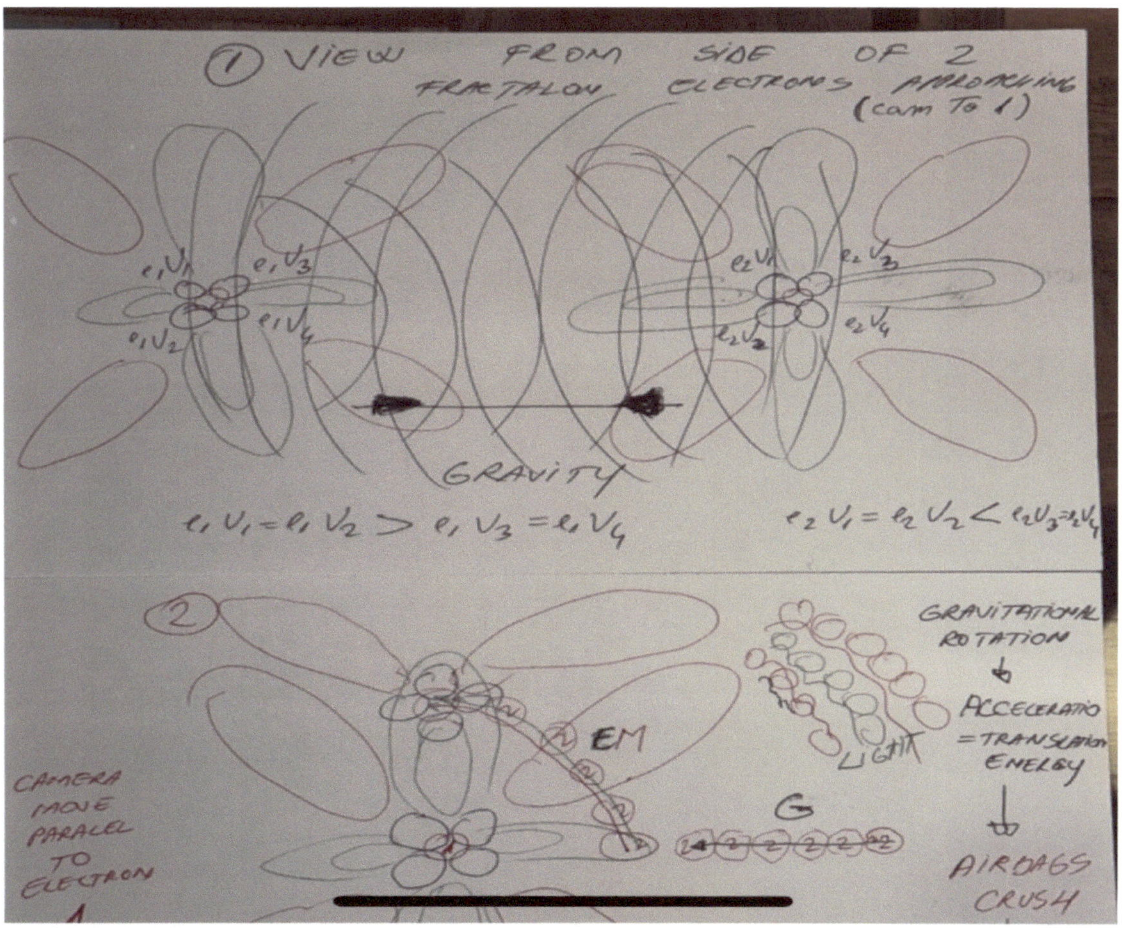

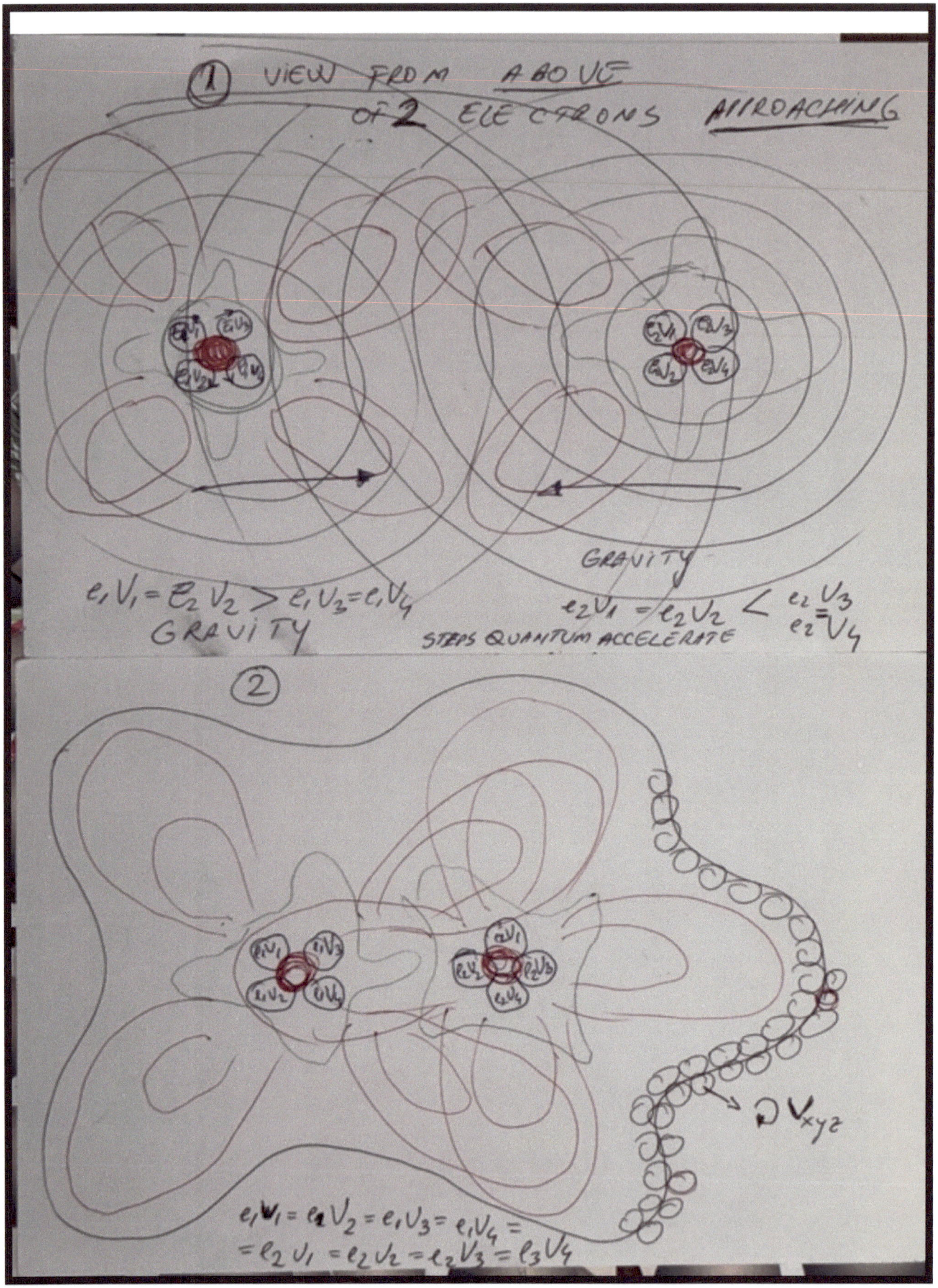
① VIEW FROM ABOVE
OF 2 ELECTRONS APPROACHING
GRAVITY
$e_1V_1 = e_2V_2 > e_1V_3 = e_1V_4$
GRAVITY
$e_2V_1 = e_2V_2 < e_2V_3 = e_2V_4$
STEPS QUANTUM ACCELERATE
②
V_{xyz}
$e_1V_1 = e_1V_2 = e_1V_3 = e_1V_4 =$
$= e_2V_1 = e_2V_2 = e_2V_3 = e_3V_4$

It will successively change the position with the neighboring Higgs bosons on the direction of the first and the other will do the same, but faster and faster, they will steal rotational energy from the neighbors and will transform this energy into translational energy (gravitational acceleration) in the moment of impact this energy will be lost by quanta (with each precession a part will leave in the system) in the form of light. To understand it is volume 3 of Fractatalon Unified Theory, but it is enough to watch the Fractalon 1 video on YT or fractalon.fr

In same image the initial and final position of 2 additional Higgs's bosons, but they changed places with others.

When you add more particles, all the extra particles will become electrons because there is no more space to stay in the cubic organization, so you have to stay between 8 particles, 4 rotating in one direction and 4 in another. (Fractalon1 video from fractalon.fr) So, because of this, 2 disturbances appear in the lattice. The first is a concentric decrease in rotation causing GRAVITY and the second is a periodic translation spiral wave (the frequency depends on changes in the poles of the extra particle, you must see the video) creating ELECTROMAGNETISM .

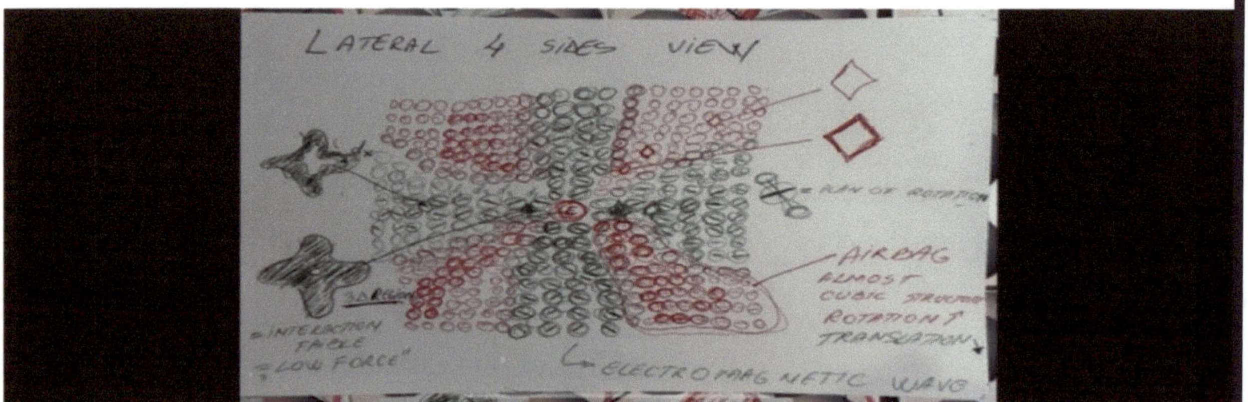

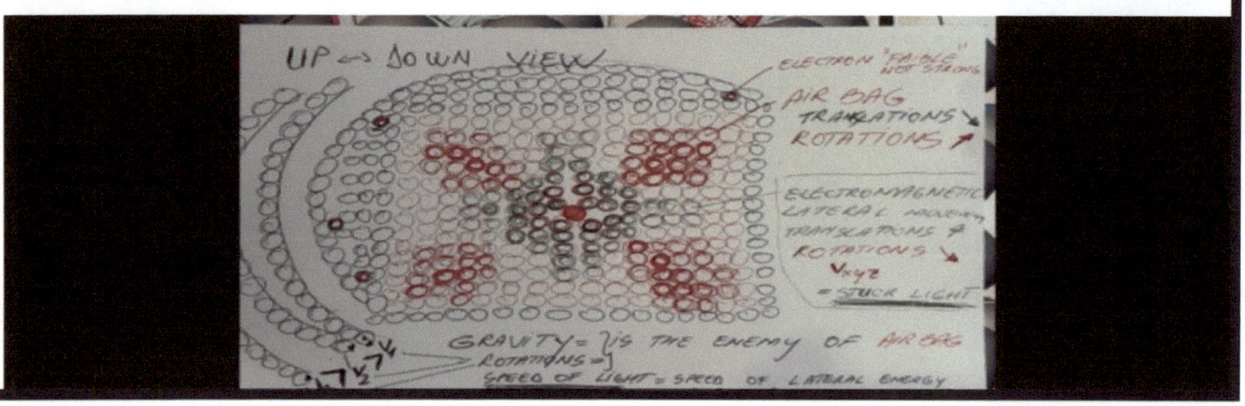

Helium atom.

Gold atom:

Simulation of change of positions between Higgs Bosons and Electrons (phenomenon present in gravitational acceleration, electric current, magnets, lightning etc.) The Higgs bosons being in a cubic network without contact with the neighbors, it can never be detected, only when it becomes electrons with precession movements. The famous wave-particle duality.

The rule for crystallization of Higgs bosons into matter, into atoms,how they will create mass and gravity .

https://youtube.com/shorts/buzDgfUaihM?si=JB0NyIM_xBKs5k-U

In the Middle Ages there was not a man in Europe who believed that the biological, not spiritual father of Jesus was a man. In the year 2025 every physicist on this planet, every physics lover on any forum still believes this religious story of forces, believes that atoms are a kind of mobile phones that can send photons for the electromagnetic force, can send gravitons for the gravitational force to every particle in the universe, and have receptors on their surface to capture gravitons and photons from all the other particles in the universe, and each particle with the super computers inside it at every moment, with systems of differential equations, Maxwell's equations, Einstein's etc. calculates its trajectory for the next moment. Not even a single person on this forum believes that each primordial particle is just a dodecahedral crystal that rotates. I don't want to offend you, but you live in the dark ages of physics and refuse to think, I lived the same way until I was 38 and then I said STOP, and in ten years of meditation I created FRACTALON UNIFIED THEORY.

The entire universe is like a giant clock. And instead of cogwheels, there are toothed balls without axis , free. When the toothed balls rotate like gyroscopes without contact with each other, in cubic structures, pole on pole, equator on equator, almost without contact, maximum possible speed, in this clock, we call this structure: THE VOID. In this perfectly symmetrical structure with balls,dodecahedrons🥎🥎 which are organized in columns, halves which turn in one direction, halves in the other, the speed of rotation because of the equatorial contact is maximum and if you manage to create a single precession, of a single ball, this energy will be transmitted with the speed of rotation of each ball which will lose the equatorial contact like a top, a gyroscope, and this speed of rotation and reorganization of the balls is "c" - THE SPEED OF LIGHT IN A VACUUM. But in this structure time does not exist because there is no contact between the balls. Each ball turns quietly without anyone knowing that it is there. On the other hand the potential to transmit a precession is maximum.

Hundreds and thousands of emails, posts with detailed explanations, videos, CONVERSATIONS OF HUNDREDS OF PAGES AND I HAVE NOT FOUND EVEN A SINGLE MAN OR PHYSICIAN WHO WILL LISTEN TO ME WHAT SHOULD I SPEAK TO BELIEVE? IT'S LIKE EXPLANATING TO A SURGEON HOW TO OPERATE ON THE BRAIN AND HE CONTINUES TO CUT OFF HEADS? I FEEL LIKE VOMITING.... I'LL STOP HERE

FRACTALOR
reactor device to produce infinite amounts of water and energy through vacuum crystallization, patented at INPI France

www.ingramcontent.com/pod-product-compliance
Lightning Source LLC
Chambersburg PA
CBHW041321210926
51847CB00002B/12
* 9 7 8 2 9 5 8 1 4 8 2 5 6 *